THE ERA OF MULTI-MESSENGER SOLAR PHYSICS

IAU SYMPOSIUM 372

COVER ILLUSTRATION:

A high resolution image of a solar plage, acquired with the VBI instrument on the Daniel K. Inouye Solar Telescope in the H-β line. The region, about 55 $\times$ 55 Mm in size, is the putative source region of the in-situ measurements obtained by Parker Solar Probe during Encounter #12. An artist impression depicts the magnetic switchbacks observed by PSP, which might be originating at the edges of supergranules and plage magnetic elements. The original H-β image was acquired on disk but has been rotated to make it appear as observed at the solar limb. (Credits: NSO/AURA/NSF, see https://nso.edu/press-release/u-s-national-science-foundation-celebrates-the-inauguration-of-its-daniel-k-inouye-solar-telescope/. Graphics by John Williams, NSO)

IAU SYMPOSIUM PROCEEDINGS SERIES

INTERNATIONAL ASTRONOMICAL UNION

UNION ASTRONOMIQUE INTERNATIONALE

International Astronomical Union

IAU

THE ERA OF MULTI-MESSENGER SOLAR PHYSICS

PROCEEDINGS OF THE 372nd SYMPOSIUM OF THE INTERNATIONAL ASTRONOMICAL UNION HELD IN BUSAN, REPUBLIC OF KOREA 2–5 AUGUST, 2022

Edited by

GIANNA CAUZZI

National Solar Observatory, USA and
National Institute for Astrophysics (INAF), Italy

and

ALEXANDRA TRITSCHLER

National Solar Observatory, USA

CAMBRIDGE
UNIVERSITY PRESS

CAMBRIDGE UNIVERSITY PRESS
University Printing House, Cambridge CB2 8BS, United Kingdom
1 Liberty Plaza, Floor 20, New York, NY 10006, USA
10 Stamford Road, Oakleigh, Melbourne 3166, Australia

First published 2023

Printed in Great Britain by Henry Ling Limited, The Dorset Press, Dorchester, DT1 1HD

Typeset in System LaTeX 2ε

A catalogue record for this book is available from the British Library Library of Congress Cataloguing in Publication data

This journal issue has been printed on FSC™-certified paper and cover board. FSC is an independent, non-governmental, not-for-profit organization established to promote the responsible management of the world's forests. Please see www.fsc.org for information.

ISBN 9781009352888 hardback
ISSN 1743-9213

Table of Contents

Preface

IAU Symposium 372, "The Era of Multi-Messenger Solar Physics" was coordinated through Division E (Sun and Heliosphere), with support from its Commissions, Division G (Stars and Stellar Physics), and the Inter-Division E-F-G Working Group on Impact of Magnetic Activity on Solar and Stellar Environments. It was held on Aug 2–5, 2022, during the first week of the XXXI IAU General Assembly in Busan, Republic of Korea. Almost 80 researchers contributed to the Symposium with invited, oral and poster presentations, representing 18 countries. Given the travel limitations due to the lingering COVID19 pandemic, over 30 contributors participated remotely in this hybrid meeting. However, as typical for GA-related Symposia, numerous other colleagues participated in selected oral and poster sessions.

Multi-messenger science has a long history in solar astronomy. For decades, direct measures of particles emanating from the Sun, like the solar wind, have been used together with electromagnetic field diagnostics as the couriers of information. This approach has shaped our understanding of how solar activity forms and relentlessly influences the heliosphere, including the planets existing within it. Several major solar facilities have recently come online, including the Parker Solar Probe (PSP), Solar Orbiter, or the Daniel K. Inouye Solar Telescope (DKIST, still in commissioning at the time of the Symposium), and they will be soon joined by others like the Chinese ASO-S or the Indian Aditya-L1. This collective, and unprecedented, observational power is now heralding an exciting new era of scientific opportunities within multi-messenger solar physics. The main goal of the Symposium was that of highlighting how these facilities can address many of the scientific questions facing contemporary Solar Physics, with a special emphasis on their synergies and the challenges and rewards of their coordinated operation.

The Symposium was structured in 7 Sessions, and included 15 invited and 24 contributed talks, as well as 36 poster and e-talk presentations. Among the many new results presented, we point out the first observations of the magnetic fields of active regions on the "hidden" face of the Sun, obtained with Solar Orbiter/PHI, and the spectacular data acquired with the Daniel K. Inouye Solar Telescope, revealing details on scales of 20–30 km at the solar surface (cf. the cover image for this book). Of note, the DKIST data were acquired during a campaign to support Parker Solar Probe encounter #12, with the main goal of providing the magnetic configuration for the source region of PSP's in-situ measurements. This perfectly introduced the topic of the Symposium. The power of combining observations of various kinds was well illustrated by modeling efforts for multi-mission science, aimed at maximizing the likelihood of linking in-situ data to their source region at or near the Sun with careful predictions.

Solar activity at a variety of spatial and temporal scales was the focus of several talks, ranging from the working of the solar dynamo as derived from a new generation of MHD simulations of the solar interior, that incorporate enhanced helioseismology techniques, to new methods to estimate the magnetic non-potentiality of active regions, which help predicting whether they will produce eruptive flares. The resulting Coronal Mass Ejections (CMEs), their magnetic configuration and interaction, as well as the propagation of solar particles in the inner heliosphere were also discussed; results from an amazing fleet of spacecrafts at our disposal are really starting to explain how the Sun shapes its environment and help our understanding of similar phenomena at other stars. Indeed, the synergy between solar and stellar studies is an emerging field; during the Symposium we discussed the long-term evolution of the solar wind, and how the study of exoplanetary systems could improve our understanding of the evolution of the solar wind and how it affected the young Earth. Of course, the proximity of our star allows us to obtain exquisitely resolved observations: these motivate sophisticated numerical investigations of mechanisms such as turbulence and magnetic reconnection, or the role of small scales in the creation of larger structures. One such a case is the local dynamo, the process responsible for the appearance of the magnetic fields in the quiet Sun and their variation with the cycle; other talks showed how small-scale vortex motions in the lower atmosphere could be the ultimate cause for the patchy nature of switchbacks in the solar wind as observed by PSP.

The Symposium concluded with the GA-plenary talk on August 5, presented by Valentin Martinez Pillet. His talk highlighted several science cases that will critically depend on the optimal coordination among multiple facilities, making a profound case for multi-messenger solar physics.

We would like to thank here the International Astronomical Union for their support, in particular for their generous policy of travel grants to young researchers. Very special thanks however are due to the Korean National Organizing Committee of IAUGA 2022 and the Local Organizing Committee, for their endless support in the organization of the whole GA, among the many logistical difficulties introduced by the pandemic. It has been truly a herculean effort, and they deserve our deepest, and heartfelt, gratitude.

Gianna Cauzzi and Alexandra Tritschler, co-chairs SOC
Boulder, CO
25 April 2023

Editors

Gianna **Cauzzi**, National Solar Observatory, USA *and* National Institute for Astrophysics (INAF), Italy
Alexandra **Tritschler**, National Solar Observatory, USA

SOC

LOC (all from Republic of Korea)

Acknowledgments

The symposium was coordinated through IAU Division E (Sun and Heliosphere) and sponsored and supported by IAU Division G (Stars and Stellar Physics), including several of their associated Commissions: E1 (Solar Radiation and Structure); E2 (Solar Activity), E3 (Solar Impact throughout the Heliosphere), as well as the Inter-Division E-F-G Working Group on Impact of Magnetic Activity on Solar and Stellar Environments. Special thanks are due to the Korean National and Local Organizing Committee of IAUGA 2022, for their endless support in the organization of the whole GA.

List of Participants

Robert **Allen**, Space Exploration Sector, Johns Hopkins Applied Physics Lab, USA
Frédéric **Auchère**, Institut d'Astrophysique Spatiale, Université Paris Saclay, France
Suresh Babu **Balaji**, Physics Dept., VIT Bhopal University, India
Dipankar **Banerjee**, ARIES, Aryabhatta Research Institute of Observational Sciences, India
Timothy **Bastian**, Science Support and Research, National Radio Astronomy Observatory, USA
Allan Sacha **Brun**, Dept. Astrophysics - AIM, CEA Paris-Saclay, France
Gianna **Cauzzi**, National Solar Observatory, USA
Jongchul **Chae**, Department of Physics and Astronomy, Seoul National University, Republic of Korea
Bin **Chen**, Dept. of Physics, New Jersey Institute of Technology, USA
Hechao **Chen**, School of Earth and Space Sciences, School of Earth and Space Sciences, China
Yajie **Chen**, School of Earth and Space Sciences, School of Earth and Space Sciences, China
Kyung-Suk **Cho**, KASI, Republic of Korea
Gwangson **Choe**, School of Space Research, Kyung Hee University, Republic of Korea
Joao M. **Da Silva Santos**, National Solar Observatory, USA
Andrea **Diercke**, National Solar Observatory, USA
Henrik **Eklund**, Department of Astronomy, Stockholm University, Sweden
Adam **Finley**, Dept. Astrophysics - AIM, CEA Paris-Saclay, France
Sarah **Gibson**, High Altitude Observatory, NCAR, USA
Tingyu **Gou**, Department of Geophysics and Planetary Sciences, University of Science and Technology of China, China
Hideyuki **Hotta**, Faculty of Science, Chiba University, Japan
Neal **Hurlburt**, Space Sciences & Instrumentation, Lockheed Martin Advanced Technology Center, USA
Hyun-Jin **Jeong**, School of Space Research, Kyung Hee University, Republic of Korea
Bibhuti Kumar **Jha**, Solar Astrophysics, ARIES Nainital, India
Bhuwan **Joshi**, Udaipur Solar Observatory, Physical Research Laboratory, India
Devojyoti **Kansabanik**, National Centre for Radio Astrophysics – Tata Institute of Fundamental Research, Pune, India
Yukio **Katsukawa**, National Astronomical Observatory of Japan, Japan
Elena **Khomenko**, Instituto de Astrofisica de Canarias, Spain
Roksoon **Kim**, Space Science Division, Korea Astronomy and Space Science Institute (KASI), Republic of Korea
Pankaj **Kumar**, American University/NASA GSFC, USA
Hannah **Kwak**, Department of Physics and Astronomy, Seoul National University, Republic of Korea
Bendict **Lawrance**, Dept. of Astronomy and Space Science, Kyung Hee University, Republic of Korea
Harim **Lee**, Dept. of Astronomy and Space Science, Kyung Hee University, Republic of Korea
Jin-Yi **Lee**, Dept. of Astronomy and Space Science, Kyung Hee University, Republic of Korea
Kyoung Sun **Lee**, Astronomy program, Department of Physics and Astronomy, Seoul National University, Republic of Korea
Ting **Li**, Research Group of Solar Magnetic Activities, National Astronomical Observatories, Chinese Academy of Sciences, China
Daye **Lim**, Department of Astronomy and Space Science, Kyung Hee University, Republic of Korea
Eun-Kyung **Lim**, Space Science Division, KASI, Republic of Korea
Yukun **Luo**, Beihang University, Beijing, China
Bhupendra **Malvi**, Barkatullah University, Bhopal, Madhya Pradesh, India
Juan Carlos **Martinez Oliveros**, Space Sciences Laboratory, University of California Berkeley, USA

Valentin **Martinez Pillet**, National Solar Observatory, USA
Christian **Moestl**, Space Research Institute, Austrian Academy of Sciences, Austria
Yong-Jae **Moon**, School of Space Research, Kyung Hee University, Republic of Korea
Divya **Oberoi**, National Centre for Radio Astrophysics, Tata Institute of Fundamental Research, India
Sneha **Pandit**, Institute of Theoretical Astrophysics, Rosseland Centre for solar Physics, Norway
Vemareddy **Panditi**, Sun and Solar System, Indian Institute of Astrophysics, India
Jinhye **Park**, Dept. of Astronomy and Space Science, Kyung Hee University, Republic of Korea
Sumiaya **Rahman**, School of Space Research, Kyung Hee University, Republic of Korea
Jack **Reid**, School of Mathematics & Statistics, University of St. Andrews, UK
Thomas **Rimmele**, National Solar Observatory, USA
Marco **Romoli**, Dept. of Physics and Astronomy, Università di Firenze, Italy
Maryam **Saberi**, Institute of Theoretical Astrophysics, Rosseland Centre for Solar Physics, Norway
Clementina **Sasso**, Osservatorio Astronomico di Capodimonte, INAF, Italy
Arpit **Shrivastav**, Aryabhatta Research Institute of Observational Sciences, India
Sami **Solanki**, MPI for Solar-System Research, Germany
Jihyeon **Son**, School of Space Research, Kyung Hee University, Republic of Korea
Anu B **Sreedevi**, Department of Physics, Indian Institute of Technology (BHU), India
Hanna **Strecker**, Solar System Department, Instituto de Astrofísica de Andalucía, Spain
Yang **Su**, Key Laboratory of Dark Matter and Space Astronomy, Purple Mountain Observatory, Chinese Academy of Sciences (CAS), China
Durgesh **Tripathi**, Inter University Centre for Astronomy and Astrophysics, India
Alexandra **Tritschler**, National Solar Observatory, USA
Vishal **Upendran**, Inter University Centre for Astronomy and Astrophysics, India
Ilya **Usoskin**, Faculty of Science, University of Oulu, Finland
Santiago **Vargas Domínguez**, Observatorio Astronomico Nacional, Universidad Nacional de Colombia, Colombia
Aline **Vidotto**, Leiden Observatory, Leiden University, The Netherlands
Arturs **Vrublevskis**, Ventspils University of Applied Sciences, Latvia
Andrew **Walsh**, ESAC, European Space Agency
Ruihui **Wang**, Beihang University, Beijing, China
Haruka **Washinoue**, University of Tokyo, Japan
Yu **Xu**, Peking University, China
Zihao **Yang**, School of Earth and Space Sciences, Peking University, China
Sibaek **Yi**, School of Space Research, Kyung Hee University, Republic of Korea
Zebin **Zhang**, Beihang University, Beijing, China

Chapter I:
NEW SOLAR FACILITIES AND MULTI-MISSION SCIENCE

NSO (USA) Director Valentin Martinez Pillet delivers the Plenary Talk of IAU S372, on August 5, 2022.

The Era of Multi-Messenger Solar Physics
Proceedings IAU Symposium No. 372, 2023
G. Cauzzi & A. Tritschler, eds.
doi:10.1017/S1743921323001266

Solar Physics in the 2020s: DKIST, Parker Solar Probe, and Solar Orbiter as a Multi-Messenger Constellation

V. Martinez Pillet[1], G. Cauzzi[1], A. Tritschler[1], L. Harra[2], V. Andretta[3], A. Vourlidas[4], N. Raouafi[4], B. L. Alterman[5], L. Bellot Rubio[6], S. R. Cranmer[7], S. Gibson[8], A. De Groof[9], Y.-K. Ko[10], S. T. Lepri[11], J. Linker[12], D. M. Malaspina[7], S. Matthews[13], D. Müller[14], S. Parenti[15], G. Petrie[1], D. Spadaro[16], I. Ugarte-Urra[10], H. Warren[10] and I. Zouganelis[9]

[1]National Solar Observatory, Boulder, CO, USA

[2] Physikalisch-Meteorologisches Observatorium Davos/World Radiation Center & ETH-Zürich, Switzerland

[3]INAF/OACN, Italy

[4] Johns Hopkins University Applied Physics Lab, MD, USA

[5] Space Science and Engineering, Southwest Research Institute, San Antonio, TX, USA

[6]Instituto de Astrofísica de Andalucia, Granada, Spain

[7]Department of Astrophysical and Planetary Sciences, Laboratory for Atmospheric and Space Physics, University of Colorado, Boulder, CO, USA

[8]High Altitude Observatory, NCAR, Boulder, CO, USA

[9]European Space Agency, European Space Astronomy Centre, Madrid, Spain

[10]Space Science Division, US Naval Research Laboratory, Washington, D.C., USA

[11]Department of Climate and Space Sciences and Engineering, University of Michigan, Ann Arbor, MI, USA

[12]Predictive Science Inc., San Diego, CA, USA

[13]UCL Mullard Space Science Laboratory, UK

[14]European Space Agency, European Space Research and Technology Centre, Noordwijk, The Netherlands

[15]Université Paris-Saclay, CNRS, Institut d'astrophysique spatiale, 91405, Orsay, France

[16]INAF/OACT, Italy

Abstract.
The National Science Foundation (NSF) Daniel K. Inouye Solar Telescope (DKIST) has started operations at the summit of Haleakalā (Hawai'i). DKIST joins the nominal science phases of the NASA and ESA Parker Solar Probe and Solar Orbiter encounter missions. By combining in-situ measurements of the near-Sun plasma environment and detailed remote observations of multiple layers of the Sun, the three observatories form an unprecedented multi-messenger constellation to study the magnetic connectivity in the solar system. This work outlines the synergistic science that this multi-messenger suite enables.

Keywords. Sun: general, Sun: atmosphere, Sun: magnetic fields, (Sun:) solar wind

---

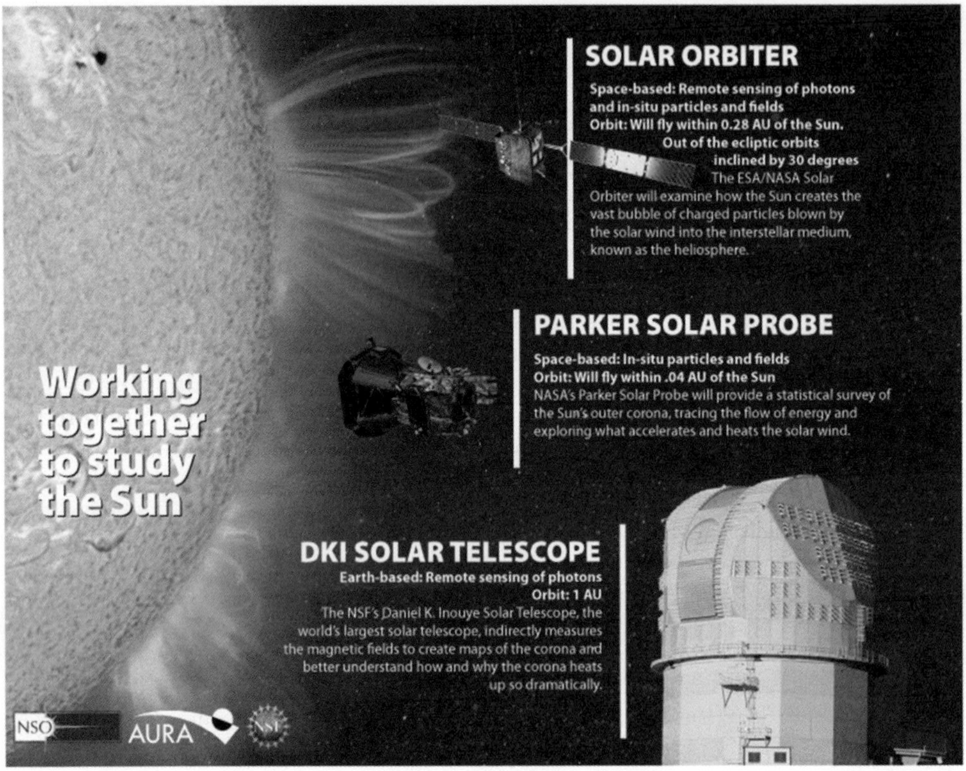

Figure 1. DKIST, PSP, and Solar Orbiter form a multi-messenger campaign that will revolutionize coronal and heliospheric physics over the next decade.

1. The multi-messenger constellation

This decade sees the start of the NSF's Daniel K. Inouye Solar Telescope (DKIST, Rimmele et al. 2020) operations coinciding with the first encounter missions aimed at mapping the physical conditions in the vicinity of the Sun, the NASA mission Parker Solar Probe (PSP, Fox et al. 2016, launched in August 2018) and the ESA/NASA mission Solar Orbiter (Müller et al. 2020, launched in February 2020). Joint observations between DKIST, PSP, and Solar Orbiter (Figure 1) will provide an unprecedented opportunity to study the solar corona and inner heliosphere, with the final aim of understanding how stars create and control their magnetic environments. NSF's DKIST contributes in many ways to this effort. In particular, a fundamental aspect of the synergistic science is the ability of DKIST to measure the off-limb solar corona and map its magnetic field quantitatively.

The PSP and Solar Orbiter in-situ instruments measure the plasma particles' kinetic properties, ion composition and elemental abundances, and local electric and magnetic fields. They also have the ability to remote-sense solar radio emission, including source location determination via direction-finding analysis. DKIST and Solar Orbiter perform imaging, spectroscopy, and polarimetry of the solar atmosphere, covering the photosphere, chromosphere, and the corona. Both PSP and Solar Orbiter image the tenuous corona and heliosphere they fly through using imagers mounted on the side of the spacecraft. By combining their data and investigating the cause-effect relationship between the processes observed in the solar atmosphere and their consequences in the solar system, the three observatories effectively create a multi-messenger system for solar and heliospheric physics.

This paper highlights the science enabled by the DKIST's new capabilities in conjunction with PSP and Solar Orbiter. The National Solar Observatory (NSO), and other ground-based observatories, also contribute to this multi-messenger era with data from synoptic observations such as those performed by GONG (Hill 2018) or the Mauna Loa Solar Observatory coronagraphs, K-Cor and (U)CoMP (Tomczyk et al. 2008). (See Schwadron et al.† for a complete discussion of ground-based support for PSP.) Of particular relevance to both space missions are synoptic magnetograms produced by GONG and SDO/HMI that help predict the instantaneous magnetic connectivity of the spacecraft during the encounter periods—modeling of the PSP connectivity during its first perihelion is discussed in van der Holst et al. (2019) and Riley et al. (2019); for the Solar Orbiter mission see Rouillard et al. (2020).

2. Tracing the Magnetic Connectivity from the Sun into the Heliosphere

Understanding the magnetic connectivity between the planets and the Sun, and by extension between exoplanets and their parent stars, requires tracing particles and fields measured in the heliosphere back to their source regions on the Sun. Establishing the underlying mechanisms responsible for this connectivity has been a daunting task from 1 AU. The exceptions are (i) major Coronal Mass Ejections (CMEs) whose progenitors at the solar surface are often well-identified thanks to the association of CMEs with flares, dimming effects, and the in-situ detection of the resulting ejected cloud (Gopalswamy et al. 2018), and (ii) relativistic electron beams, which can, for some events, be traced back to their acceleration location in the corona via hard X-ray and radio emission (e.g., Cairns et al. 2018; Krucker et al. 2008). Other, steadier and subtler processes dictating this connectivity, such as the various forms of the solar wind, have not been conclusively linked with processes and evolving configurations on the Sun. The fast solar wind is known to originate in open field regions or coronal holes, and long-lived low-latitude coronal holes are known to drive recurring high-speed solar wind streams at the Earth (e.g., Gibson et al. 2009). However, the specific small-scale structures (flux tubes, etc.) from where it emanates and the underlying physics remain elusive (Cranmer & Winebarger 2019). The mapping of the slow solar wind into specific solar structures is also poorly understood. Its creation is attributed to a variety of configurations, such as the heliospheric current sheet and pseudostreamers (Abbo et al. 2016). Beyond this, our knowledge about the exact location of the processes that give rise to the expanding outer atmosphere of the Sun (see Zhao et al. 2017, and references therein) is rather inconclusive.

At 1 AU, plasma from the Sun has evolved and mixed with other constituents via transport and interactions resulting in conditions that prevent a conclusive mapping to its solar sources. Similarly, the absence of measurements of the solar wind close to the Sun has challenged the validation of competing models for its acceleration. Near-Sun solar wind samples should be easier to map with sufficient precision to a solar source region and a specific process. For this reason, over several decades, the solar community has advocated for missions that approach our star and measure the fields and particles closer while they still preserve many of the properties (composition, charge, kinetic properties, etc.) that link them to their solar sources. This proximity to the Sun was the rationale behind the Parker Solar Probe (Fox et al. 2016) mission in the NASA Heliophysics

† `https://sppgway.jhuapl.edu/sites/default/files/Pubs/SPP-GBN-WhitePaper-v5.0.pdf`

portfolio and the Solar Orbiter mission (Müller et al. 2020) included in the European Space Agency (ESA) science program also participated by NASA.

PSP and Solar Orbiter are encounter-class missions instead of observatory-class missions. No formal calls for observing proposals are submitted to the mission teams. Instead, predefined measurements are made during selected windows coinciding with the most scientifically relevant portions of the orbits, in particular the periods of closest proximity to the Sun and largest inclination out of the ecliptic plane.

Both missions use Venus Gravity Assist Maneuvers (GAM) to get closer to the Sun (PSP) and, in the case of Solar Orbiter, to increase the angle to the plane of the ecliptic, allowing a better view of the solar poles and mapping of the 3D heliospheric volume. A mission like Solar Orbiter that escapes the ecliptic to provide imagery of the solar poles can make a profound impact on our understanding of the solar wind formation and acceleration.

The proximity to the Sun achieved by PSP generates corotation opportunities (or quasi-corotation for Solar Orbiter) periods where the spacecraft flies right above the same solar feature for extended periods (hours and days), allowing the study of its evolution and how the properties of the solar wind vary depending on changes in the source regions. These corotation periods help disentangle spatial from temporal variations in the heliospheric structures and establish cause-and-effect relationships between solar processes and in-situ measurements.

As the encounter windows depend on the particular orbit of PSP or Solar Orbiter, they result in varying Earth-spacecraft configurations. For DKIST, the best times for synergistic science occur when Earth is in quadrature with one of the spacecrafts. In that case, off-limb coronal DKIST observations map the inner solar vicinity of the spacecraft's trajectory. However, encounter windows in alignment between the Earth and the satellites allow for additional scientific opportunities, with DKIST observing on the disk or in the off-limb corona depending on the science case.

3. PSP and DKIST: Enabling Science by Combining In-Situ Near-Sun Measurements and 1 AU Coronal Observations

PSP was launched in August 2018 and has already performed 13 perihelia at a distance of 12 $R_\odot$ (0.06 AU), following five Venus GAM. The nominal mission uses a total of 24 perihelion encounters, with the last one occurring in June 2025. This period includes about 3.5 years after the start of DKIST operations in February 2022. In this period, two additional Venus flybys will bring PSP ever closer to the Sun, reaching 9.86 $R_\odot$ (0.0459 AU) for the last three encounters in 2024/25. PSP will spend about a thousand hours below 20 $R_\odot$, and fifteen hours below 10 $R_\odot$. This inner region is crucial for the scientific objectives of PSP as at this distance, the solar wind is predicted to be sub-Alfvénic, and plasma interactions generate waves that can carry energy back towards the Sun. The coexistence of inward and outward propagating Alfvén waves drives turbulence, which in turn is often invoked as a mechanism to heat the solar corona (Cranmer & van Ballegooijen 2005). When the solar wind accelerates and becomes super-Alfvénic, the inwardly directed waves no longer reach back to the corona, and this decoupling is often taken as the definition for the initiation of the heliospheric solar wind. PSP first entered this critical region where the solar wind emanates during encounter 8 (Kasper et al. 2021) for five hours.

PSP carries three in-situ instrument suites and a heliospheric imager. The Solar Wind Electrons Alphas and Protons (SWEAP) investigation (Kasper et al. 2016) measures the particles that constitute the bulk of the solar wind: electrons, protons, and alpha particles. The four SWEAP sensors measure the velocity distribution functions of the particles with good energy and angular resolution. SWEAP is a critical instrument for connectivity studies as it detects particles that one can trace back to chemical composition on the solar atmosphere. Additional particle measurements at higher energies are performed by the Integrated Science Investigation of the Sun (IS⊙IS, Manchester et al. 2004). IS⊙IS is the primary instrument to study the particles created by energetic events, such as flares and CMEs.

The FIELDS instrument (Bale et al. 2016) measures DC-coupled magnetic and electric fields, plasma wave magnetic and electric fields, and solar radio emissions. From the measured local electric and magnetic fields, it produces an estimate of the Poynting flux, i.e., the electromagnetic energy flux crossing that region of space. The 3-components of the local magnetic field obtained by FIELDS allow mapping back to the outer corona and solar surface.

Finally, the Wide-Field Imager (WISPR, Vourlidas et al. 2016) is the only imaging instrument aboard the spacecraft. It measures the white-light scattered by dust (F-Corona) and coronal electrons (Thomson scattering). WISPR looks at the large-scale structure of the corona and solar wind before the spacecraft passes through it. It images the fine-scale structuring of the solar corona and has detected the long-predicted dust-free zone that exists near the Sun (Stenborg et al. 2022). At closest approach (in 2024), the inner field-of-view of WISPR will image about 2 solar radii above the surface, which is close to the height observable by DKIST and its instrumentation that reach to 1.5 radii. This creates a unique opportunity to test whether the diffuse coronal brightness seen in He I 1083 nm can be due to the neutralization of alpha particles by dust as proposed by Moise et al. (2010) and more recently investigated by Rivera et al. (2020) and Del Zanna et al. (2020). This is a prime example of synergistic science, where progress has been difficult thus far, but that can be enabled by coordinated observations from the DKIST's Cryogenic Near IR Spectro-Polarimeter (Cryo-NIRSP), WISPR, and SWEAP.

Clarifying the detection of neutral Helium in the million-degree solar corona is a key goal for DKIST because the He I 1083 nm line is the only permitted IR line available, and it has the potential to unlock new coronal magnetic field diagnostics through the Hanle effect as demonstrated by Dima et al. (2016). These authors describe an algorithm that combines linear polarization measurements of the Si X 1430 nm forbidden line, in the saturated Hanle regime for coronal conditions, with polarization observations of the He I 1083 nm permitted line to infer the three components of the coronal magnetic field. Such measurements will constrain the magnetic topology significantly and can feedback into global coronal models, improving the accuracy of the heliospheric magnetic configurations predicted by them. While at solar minimum, coronal-heliospheric models using photospheric synoptic data can produce satisfactory mappings of the connectivity back to the Sun, the more complex corona that PSP will encounter during solar maximum will need the most refined available input data (Raouafi et al. 2016). With this in mind, the DKIST Critical Science Plan (Rast et al. 2021) is considering a synoptic program that regularly measures the solar corona in anticipation of PSP encounters. The lines and techniques used by this program can change as we learn how to generate the best input coronal magnetic data for the models. Most likely, such a synoptic program will initially use the forbidden Fe XIII line at 1075 nm in a way similar to the CoMP instrument, but will also include the line-of-sight component of the magnetic field from the Stokes V accessible to DKIST. A first set of coronal observations of this kind was acquired on June 1st, 2022, during quadrature with PSP, just before its 12th perihelion passage.

The chemical composition of the solar wind is an indicator of the source region on the Sun (Geiss et al. 1995) and is a critical ingredient for establishing the magnetic connectivity. An example relevant to PSP is the relation between in-situ measurements of Helium at 1 AU and their dependence on the solar wind speed and phase of the activity cycle (Kasper et al. 2007, 2012). Observations show that the smallest Helium abundances are observed during sunspot minimum, with slower solar winds showing a more pronounced correlation than faster winds. In fact the Helium abundance of the fast solar wind minimally changes as a function of the cycle. These correlations point to a mechanism that affects the constituent with the highest ionization potential of the solar plasma that varies with the magnetic activity levels. The exact mechanism remains unknown, but there are indications that the highest depletion of Helium occurs when the speed of the solar wind is the slowest (about 250 km/s), indicating a trade-off between destroying Helium and the ability to accelerate the solar wind (see Alterman & Kasper 2019, and references therein). The combination of PSP and DKIST over the next decade will provide new ways to investigate this fundamental problem. Using the SWEAP instrument, PSP will measure the in-situ abundance of Helium much closer to the Sun, providing a less ambiguous identification of the source regions. As the Helium depletion is known to occur lower in the corona (Laming 2015), chromospheric observations will play a crucial role. All DKIST instruments can contribute to this investigation. DKIST can observe on the disk and off-limb several Helium lines, most importantly He I 1083 nm with the Diffraction-Limited Near-Infrared Spectro-polarimeter (DL-NIRSP) (near limb corona) and Cryo-NIRSP up to 1.5 $R_\odot$. In the visible, the Visible Spectro-Polarimeter (ViSP) can access a number of He I and He II spectral lines and observe them in the near-limb region, benefiting from the coronagraphic capabilities of DKIST and the dark sky conditions on Haleakalā. Most likely the relevance of these measurements will become more critical at the closest PSP distances near solar maximum in 2024, but the solar cycle fluctuations described above indicate that synoptic measurements should start as soon as possible, in the raising phase of the solar cycle. A first test of on-disk observations of the PSP source region, using the Visible Broadband Imager (VBI) and Visible SpectroPolarimeter (ViSP) on DKIST was performed on June 2nd and 3rd, 2022, during its first phase of operations (the Operations Commissioning Phase, OCP).

4. Solar Orbiter, PSP, and DKIST: A Multi-Messenger Approach to Solve the Magnetic Connectivity Conundrum

Solar Orbiter launched in February 2020 on a NASA Atlas V 411 rocket from Space Launch Complex 41 at Cape Canaveral. The joint ESA/NASA mission combines four in-situ instruments and six remote sensing telescopes to connect observed processes on the Sun with local measurements. Solar Orbiter seeks to understand the magnetic connectivity between the Sun and the heliosphere. As such, it was designed and built as a multi-messenger mission from its conception. With openings in the heat shield to allow the onboard telescopes to observe the Sun, it does not approach the Sun as close as PSP does. Solar Orbiter reaches a distance to the Sun of 60 $R_\odot$ (0.28 AU), and an inclination of 33° relative to the ecliptic plane at the end of its (extended) mission. It uses a series of 8+8 (nominal plus extended) encounter periods, each comprised of three 10-day windows during which the remote sensing instruments operate. In-situ instruments are always operational. The remote sensing windows are centered on the most scientifically relevant parts of the orbit (see Figure 2), perihelion, and maximum and minimum heliographic latitude, both providing vantage points for observing the solar poles. The exact timing of these windows is based on long term planning and may evolve during the mission. The choice of remote sensing windows will also consider coordinating opportunities with PSP. The first opportunity for collaboration among the three facilities occurred in June 2022, during the 12th PSP perihelion passage. The observing proposal submitted for consideration for the coordination used Director's Discretionary Time (DDT); as such, the data has no proprietary period, and will be made public soon after calibration.

The Solar Orbiter payload is more extensive than the PSP payload. All remote sensing instruments offer multiple opportunities to coordinate with DKIST. The Solar Orbiter Polarimetric and Helioseismic Imager (PHI; Solanki et al. 2020), together with

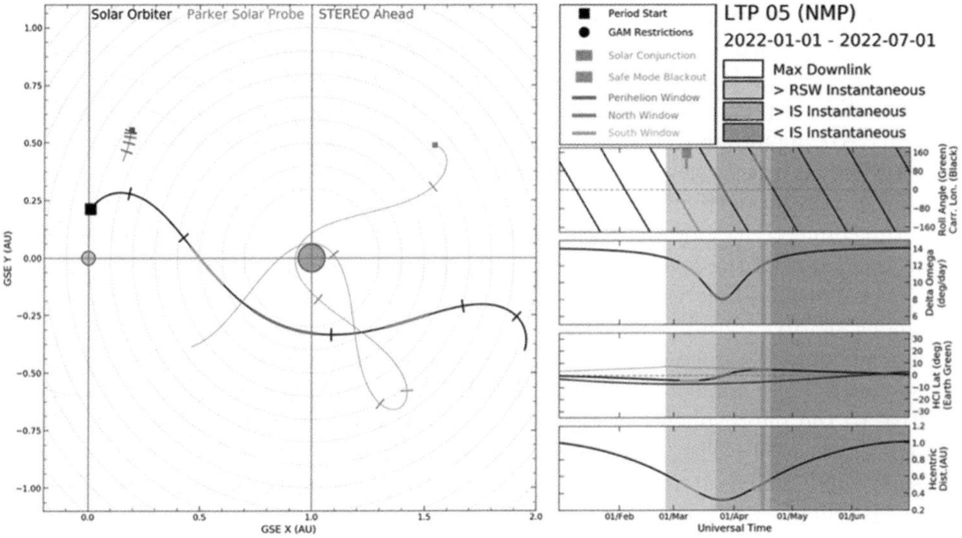

Figure 2. Sample encounter orbit for the Solar Orbiter mission. The three remote sensing windows are indicated by colors (orange, dark red, blue). The orange window corresponds to a conjunction configuration and the dark red window to a quadrature encounter. Earth is at coordinates [0,0] on the left image, Sun at [1,0]. Image courtesy of ESA.

DKIST/ViSP observations of the same spectral line, will provide an opportunity to carry out magnetic field stereoscopy for the first time and solve the 180° ambiguity intrinsic to all Zeeman measurements. This stereoscopic study will certainly include other instruments on the Earth-Sun line-of-sight, but the sensitivity to transverse fields that DKIST provides will prove crucial for satisfactory disambiguation, in particular in the quiet Sun. The Extreme UV Imager (EUI, Rochus et al. 2020) and the Spectral Imager of the Coronal Environment (SPICE Consortium et al. 2020) spectrograph, can complement the coronal measurements of DKIST in various ways depending on the spacecraft-Earth configuration during the encounters. Quadrature windows can help disentangle line-of-sight effects from the DKIST Cryo-NIRSP coronal measurements (see Judge et al. 2013) as EUI will provide the missing context of the structures contributing to the line-of-sight average. Using the Cryo-NIRSP instrument in combination with SPICE will enable for the first time stereo-spectroscopy of outflows and plasma diagnostics, which are also heavily affected by line-of-sight effects. For these synergistic examples, field extrapolations from PHI can provide quantitative information of the magnetic fields contributing to the line-of-sight average.

Solar Orbiter conjunction configurations will provide an opportunity for simultaneous on-disk coordinated observations combining the Solar Orbiter remote sensing telescopes and the DKIST on-disk instrument suite, maximizing spectral coverage (from UV to the IR), temporal and spatial resolution, and sensitivity. Also, in conjunction, off-limb observations, including the coronagraph onboard Solar Orbiter (METIS; Antonucci et al. 2020), will benefit from coordinated observations with DKIST. Specifically, DKIST direct measurements of the magnetic field in the off-limb corona provide a unique opportunity to calibrate the indirect inferences of the closed vs. open magnetic configuration planned by METIS.

The in-situ payload for Solar Orbiter includes a complete suite of instruments with analogs to those onboard PSP, but with some additional capabilities. The list consists of an energetic particle detector (EPD; Rodríguez-Pacheco et al. 2020), a magnetometer

(MAG; Horbury et al. 2020), a radio and plasma wave sensor (RPW; Maksimovic et al. 2020), and a solar wind analyzer (SWA; Owen et al. 2020). This last instrument is similar to the PSP's SWEAP suite, but with an added—NASA provided—key component, the Heavy Ion Sensor (HIS). SWA will give the solar wind composition measurements between 0.3 and 1 AU, including electrons, protons, alpha particles, but with HIS extending the composition analysis to heavier ions such as C, N, O, Si, Fe, S, and others. Measurements from SWA/HIS, in particular in combination with SPICE on-disk data, can help narrow down the connectivity options back to the Sun, as the charge-state ratios of heavy ions are an indication of the level of heating at the source and help distinguish among source candidates (coronal hole vs. streamer), although some modeling might be needed (e.g., Laming et al. 2019; Oran et al. 2017).

A key strategy to trace plasma parcels from the Sun into the heliosphere is to track the so-called First Ionization Potential (FIP)-bias, the anomalously high abundance (relative to the photosphere) of elements with low First Ionization Potential (see, e.g., Schmelz et al. 2012). This effect is more pronounced in the slow solar wind than in the fast solar wind, which often shows photospheric abundances (Lee et al. 2015). The nature of the FIP effect is still under debate, but the first ionization of neutral elements occurs in the chromosphere, a layer that DKIST will map with unprecedented detail. Indeed, it is thought that new active regions emerge with photospheric abundances and develop the FIP-bias at chromospheric heights that then propagate into the corona (Laming 2015). This processing and propagation to the upper layers can give rise to a lag between magnetic activity indices and abundance fluctuations (Alterman & Kasper 2019). Comparing elemental abundances and FIP-bias inferred from remote sensing observations with in-situ counterparts is the main objective of the SPICE and SWA/HIS combination. Both instruments can determine the strength of the FIP-bias for several chemical elements. Coordinated observations tracking the same plasma packages are planned for several perihelion windows. This intrinsically multi-messenger approach to study the magnetic connectivity between the Sun and the heliosphere is one of the overarching goals of the Solar Orbiter mission. As is the case for PSP, by approaching the Sun to about 0.3 AU, Solar Orbiter will certainly facilitate tracing the plasma parcels back to the surface. DKIST can provide invaluable support by observing various layers of the Sun with different instruments and in distinct orbital configurations. It is now clear that the wavelength range covered by Cryo-NIRSP contains coronal lines which are potentially very unique for the FIP bias study (for instance, Ar XIII 834nm & 1014nm, Si IX 2.58μm & 3.93μm, S XIII 1030nm, S XI 1.39μm & 1.92μm, Fe XIII 1.07μm & 1.08μm; see Del Zanna & DeLuca 2018). The inclusion of additional filters that observe such lines will complement the temperatures and elemental abundances coverage of the FIP bias studies from SPICE. Other visible lines from the ViSP instrument can provide additional diagnostics, but the coronal capabilities of this instrument need to be demonstrated first.

One area where DKIST can critically contribute to this investigation is by providing accurate magnetic field measurements during the period when Solar Orbiter is establishing the connectivity between the SPICE composition measurements and the SWA/HIS in-situ measurements. For orbital settings where the SPICE FOV is on the disk, DKIST can co-point and obtain vector magnetic maps in the photosphere and chromosphere. By adding the chromospheric magnetic information, DKIST can map the layer where composition (FIP bias) changes and where field line reconfiguration is expected to occur. A key target is a process known as interchange reconnection (Crooker et al. 2002; Fisk & Schwadron 2001) whereby a closed field region interacts and reconnects with an open field line, establishing a new route for plasma to escape the Sun. The FIP bias occurs in the closed loop before interchange reconnection. After interchange reconnection, the FIP biased plasma is transmitted via the newly open channel into the solar wind (for a recent discussion of the factors at play, see Laming et al. 2019). It is crucial to understand if this reconnection is one of the main processes that give rise to the slow solar wind, and its identification and characterization require careful measurements of the changing connectivity. In quadrature encounters, DKIST will similarly seek to quantify the degree of field line openings occurring in coronal levels and the expansion rate of the resulting open field lines. We anticipate targeting both the near-limb region (DL-NIRSP) and the large-scale corona far from the limb (Cryo-NIRSP).

The coordinated observations described above have various potential target regions such as small-scale emergence episodes in the interior of coronal holes (e.g., Shimojo & Tsuneta 2009) or the dynamics near their boundaries, as both have been consistently related with the slow solar wind (Ko et al. 2014). After the discovery of ubiquitous upflows near the edges of active regions by Hinode (Harra et al. 2008), these areas may constitute a source of the nascent slow solar wind. Using Hinode-estimated FIP bias ratios and in-situ composition data at 1 AU, Brooks & Warren (2011) traced back slow wind-like composition measurements to the edges of an AR that displayed outflows for at least five days. Thus, active-region boundaries represent another obvious target for coordinated observations between Solar Orbiter, Hinode, and DKIST. The additional out of the ecliptic vantage point provided by Solar Orbiter will produce new insights into the solar wind's source regions (see also the recent work by Brooks et al. 2022).

Coordination, including all three facilities, DKIST, PSP, and Solar Orbiter, also offers unique opportunities for synergistic science. The previous section already explained the case for combining remote sensing and in-situ Helium observations between DKIST and PSP. Bringing Solar Orbiter into the mix can prove key understanding to the role of Helium in the various types of solar wind and the source regions. By using the EUI He II 304 Å channel in conjunction with Lyman-alpha observations from EUI and METIS, Solar Orbiter will be able to estimate the Helium abundance during the remote sensing windows. Coordinated observations with DKIST during on disk and off-limb passages using He I 1083 nm (and potentially other lines in the visible with the ViSP) will allow tracking the He (the highest FIP element) on the Sun from even deeper layers than what Solar Orbiter can do on its own. Combining DKIST and Solar Orbiter data during conjunction windows to map the presence of Helium in the solar corona while PSP measures in quadrature alpha particles from within 10 $R_{\odot}$ is a compelling way to understand the evolution of Helium in the nascent solar wind. It remains to be seen if the orbital mechanics of both missions will make possible such a joint campaign, in particular, towards the end of the PSP prime mission phase.

5. DKIST, PSP, Solar Orbiter, and spacecraft at 1 AU: Studying the evolution of CMEs observed in conjunction.

CME properties evolve as they propagate through the solar wind due to expansion and interaction with the solar wind (e.g., Bothmer & Schwenn 1996; Manchester et al. 2004; Lavraud et al. 2014; Ruffenach et al. 2015). A complete understanding of the changes in the properties of CMEs as they propagate in the inner heliosphere is required to understand their space weather impact at the Earth. Prior to PSP, the most recent in situ magnetic field and plasma observations of CMEs at <0.7 AU came from the MESSENGER spacecraft at Mercury (e.g., Good et al. 2015; Winslow et al. 2016). However, MESSENGER was a planetary mission, and although its observations were useful for heliophysics studies, they were quite limited. With the era of DKIST, PSP, and Solar Orbiter taking unprecedented measurements of the Sun and inner heliosphere environment, time is ripe to develop our physical understanding of the evolution and interaction of CMEs prior to reaching Earth, and thereby lay the groundwork for improved space weather forecasting. These three facilities will allow for unparalleled mapping of CMEs as they propagate from the Sun to 1 AU. For CMEs that are observed in conjunction, this can be achieved by DKIST providing accurate magnetic field measurements of the source as the CME erupts, PSP and/or Solar Orbiter taking in situ magnetic field and plasma measurements of the CME inside 1 AU, and STEREO/ACE/Wind/IMAP providing in situ measurements at 1 AU.

Even better monitoring will be achieved for CMEs that are observed in conjunction by both PSP and Solar Orbiter, providing two observations points in the innermost heliosphere, but in quadrature with respect to the Earth. This configuration allows the novel coronal DKIST observations to map the prominence-cavity system before it erupts. Coronal prominence-cavity systems are regions of stored magnetic energy that erupt in the form of CMEs. Already in quiescent conditions, the equilibrium state of the system is only partially understood. The inner prominence shows various types of small-scale dynamics which possibly induce material exchange with the cavity itself. High temperature and low density dominate the cavity. The models proposed for the cavity lack the constraints of magnetic field measurements, and only a few measurements of prominences magnetic field have been obtained (but see Wang et al. 2020). In this respect, DKIST will be able to provide accurate and complete magnetic field measurements in both regions. The coordination with Solar Orbiter (using PHI and EUI) in quadrature will allow better constraint of the 3D magnetic structure and the changing topology from the chromosphere to the corona. This will also be possible thanks to the high latitudes reached by Solar Orbiter later in the mission, which will provide a completely new view of the cavity-prominence system. The dynamics (as well as other plasma diagnostics) of prominences and their interface with cavities will be obtained using the Doppler information from the SPICE spectra, together with DKIST in coronal mode. These studies will constrain eruption models with more quantitative and reliable information on the precursor state of the system.

6. Opportunities for Discovery

The previous sections showcased some of the combined science cases as opportunities for enhanced, synergistic science. Close coordination among the three facilities will lead to additional, often unanticipated opportunities. As a case example, DKIST, PSP, and Solar Orbiter will shed light into the so-called open flux problem (Linker et al. 2017), i.e., the persistent underestimation-by as much as a factor three, of the heliospheric magnetic flux predicted by models based on surface magnetograms, vs. what is measured in situ at 1 AU. PSP data has shown that the open flux problem persists in the inner heliosphere all the way to 0.13 AU (see Badman et al. 2021). The high resolution, high sensitivity vector magnetograms from DKIST will unveil the amount of flux that currently lies undetected in existing full-disk magnetograms, and that can be responsible for the larger flux observed at 1 AU.

Another case example is the lingering problem in solar and stellar astronomy of how stars create X-ray emitting coronae and winds. The processes leading to coronal heating are ultimately responsible for the escape of plasma from the Sun into the heliosphere. However, how the plasma is heated is still debated. A new generation of numerical models of the solar atmosphere, including MHD and radiative effects (Hansteen et al. 2015; for a review see Klimchuk 2006, 2015), have shown that the ever-present reconfiguration of chromospheric and coronal magnetic fields leads to the formation of magnetic discontinuities supporting the idea of nanoflares as the underlying process leading up to atmospheric heating. However, it is also known from high-resolution imaging and spectroscopy that the corona is permeated by MHD waves that can dissipate and heat the ambient coronal plasma (Cranmer & van Ballegooijen 2005; De Pontieu et al. 2007; Morton et al. 2021).

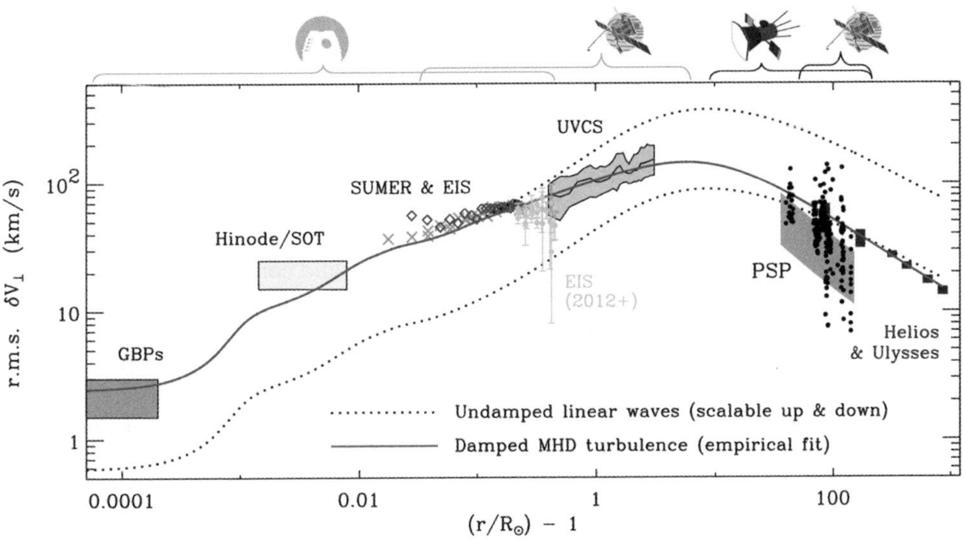

Figure 3. Velocity (rms) amplitude of transverse motions from the solar surface to the near-Earth environment. Squares and symbols inside one solar radius (above the limb) correspond to remote sensing observations as indicated. Blue squares are measurements from in-situ space missions. The solid and dashed lines correspond to damped and undamped Alfvénic wave turbulence (Cranmer & van Ballegooijen 2005) models. The upper part of this figure indicates the regions accessible to DKIST observations (and, in some cases, Solar Orbiter remote sensing instruments) and to PSP and Solar Orbiter in-situ measurements. New analysis of UVCS/SOHO H I Lyα data (green region) is discussed further by (Cranmer 2020). PSP data from the first two perihelia were analyzed by (Chen et al. 2020, light-blue region) and by S. Cranmer ($v \geq$ 450 km/s; black points), but they were dominated by slow-speed solar wind. Adapted from Cranmer et al. (2017).

It is almost certain that both processes, reconnection and wave dissipation, occur on the Sun in various forms and relevance depending on the specific region under study. Progressing in our understanding of the processes that dominate the heating and their underlying physics is a critical challenge in solar research where the multi-messenger approach represented by DKIST, PSP, and Solar Orbiter will provide new perspectives.

Figure 3 (adapted from Cranmer et al. 2017) shows a clear example of how the multi-messenger constellation can synergistically produce constraints to the various physical processes that lead to heating. The lines in the figure indicate the magnitude of the transverse motions present in a model of damped (solid) and undamped (dashed) Alfvén wave turbulence compatible with solar remote sensing and in-situ observations (Cranmer & van Ballegooijen 2005). The figure starts with the photospheric transverse motions observed in the so-called G-band bright points (GBPs). These transverse motions are thought to be responsible for shaking magnetic field lines leading to upward propagating Alfvén waves. While these GBPs motions have been measured using intensity proxies for decades, very few (magnetic) wave power estimates are available (but see Jafarzadeh et al. 2013, 2017). DKIST will perform much more accurate and quantitative measurements of these photospheric motions and their magnetic consequences, as demonstrated recently by Van Kooten & Cranmer (2017). The Hinode/Solar Optical Telescope (SOT) box in yellow indicates the amplitude of the transverse motions detected by Hinode using Ca II H chromospheric images of type II spicules (De Pontieu et al. 2007). Using DKIST's higher spatial resolution and magnetic sensitivity, we will be able to quantify the energy fluxes carried out by these waves by measuring the perturbations directly in the magnetic field lines. Critical instruments for estimating the induced velocity and the magnetic fluctuations are the DL-NIRSP and the Visible Tunable Filter (VTF). Moving away from the solar surface in Figure 3, the next set of measurements corresponds to nonthermal line widths observed by instruments in the SOHO and Hinode missions. They reach out to 1 $R_\odot$ above the surface and display mixed indications of damping processes. SOHO's UltraViolet Coronagraph Spectrometer (UVCS) (green square) shows little evidence of damping. However, more recent measurements by Hinode's Extreme-ultraviolet Imaging Spectrometer (EIS) (magenta dots in the figure) show signs of damping mechanisms acting in the corona that may contrast with the UVCS analyses. It is urgent to confirm the existence, or not, of these damping processes and consolidate our understanding of how Alfvén waves can deposit their energy and convert it to heat in this inner region of the corona. Cryo-NIRSP routine observations of spectral lines in this range of heights will undoubtedly provide the observational constraints needed to clarify the amount of damping present in this region of the solar corona. Solar Orbiter coordinated observations in conjunction windows can provide nonthermal line widths of other species and create a complete picture of the damping processes at play in the solar corona. The early years of DKIST standard operations should aim at including coordinated campaigns with Hinode/EIS to ensure cross-calibrations of past and future non-thermal doppler signals.

The models in Figure 3 reach out past the Earth and can be compared to the observed in-situ magnetic field fluctuating amplitudes detected by Helios and Ulysses. A good match with existing data (blue squares) requires the presence of wave damping in the heliosphere (solid line). The instruments onboard PSP and Solar Orbiter are going to provide much better coverage over a broader range of distances to the Sun and in a variety of solar and heliospheric structures (see dots in Fig. 3). Fine tuning of models to fit the various observations from DKIST/Solar Orbiter and the in-situ detections from PSP/Solar Orbiter will fundamentally constrain the potential physical processes behind wave heating models.

7. Concluding Remarks

This work describes the unique scientific opportunities enabled by NSF's DKIST capabilities—especially for observing the solar corona—in coordination with the near-Sun PSP and Solar Orbiter missions leading to an unprecedented multi-messenger era for Solar and Heliospheric Physics. The science cases addressed here demonstrate the need for coordination between the various teams to maximize the scientific return this unique constellation offers. This coordination is critical to ensuring a collaborative framework between the remote sensing and the in-situ communities. The nature of the two communities' measurements and the data they use are fundamentally distinct despite a common origin at the Sun. The intrinsic multi-disciplinary nature of this collaboration requires coordination by the relevant stakeholders operating these facilities.

Acknowledgements

Based on the discussions at the DKIST Critical Science Plan Workshop 4: "Joint Science with Solar Orbiter and Parker Solar Probe". JHU/APL, 13 - 15 March 2018, Laurel, MD, USA

References

Abbo, L., Ofman, L., Antiochos, S. K., et al. 2016, Space Science Reviews, 201, 55
Alterman, B. L. & Kasper, J. C. 2019, ApJ Letters, 879, L6
Antonucci, E., Romoli, M., Andretta, V., et al. 2020, A&A, 642, A10
Badman, S. T., Bale, S. D., Rouillard, A. P., et al. 2021, A&A, 650, A18
Bale, S. D., Goetz, K., Harvey, P. R., et al. 2016, Space Science Reviews, 204, 49
Bothmer, V. & Schwenn, R. 1996, Advances in Space Research, 17, 319

Brooks, D. H., Janvier, M., Baker, D., et al. 2022, arXiv e-prints, arXiv:2210.08899
Brooks, D. H. & Warren, H. P. 2011, ApJ Letters, 727, L13
Cairns, I. H., Lobzin, V. V., Donea, A., et al. 2018, Scientific Reports, 8, 1676
Chen, C. H. K., Bale, S. D., Bonnell, J. W., et al. 2020, ApJ, Supplement, 246, 53
Cranmer, S. R. 2020, ApJ, 900, 105
Cranmer, S. R., Gibson, S. E., & Riley, P. 2017, Space Science Reviews, 212, 1345
Cranmer, S. R. & van Ballegooijen, A. A. 2005, ApJ, Supplement, 156, 265
Cranmer, S. R. & Winebarger, A. R. 2019, Annual Review of Astron and Astrophys, 57, 157
Crooker, N. U., Gosling, J. T., & Kahler, S. W. 2002, Journal of Geophysical Research (Space Physics), 107, 1028
De Pontieu, B., McIntosh, S. W., Carlsson, M., et al. 2007, Science, 318, 1574
Del Zanna, G. & DeLuca, E. E. 2018, ApJ, 852, 52
Del Zanna, G., Storey, P. J., Badnell, N. R., & Andretta, V. 2020, ApJ, 898, 72
Dima, G., Kuhn, J., & Berdyugina, S. 2016, Frontiers in Astronomy and Space Sciences, 3, 13
Fisk, L. A. & Schwadron, N. A. 2001, ApJ, 560, 425
Fox, N. J., Velli, M. C., Bale, S. D., et al. 2016, Space Science Reviews, 204, 7
Geiss, J., Gloeckler, G., & von Steiger, R. 1995, Space Science Reviews, 72, 49
Gibson, S. E., Kozyra, J. U., de Toma, G., et al. 2009, Journal of Geophysical Research (Space Physics), 114, A09105
Good, S. W., Forsyth, R. J., Raines, J. M., et al. 2015, ApJ, 807, 177
Gopalswamy, N., Akiyama, S., Yashiro, S., & Xie, H. 2018, Journal of Atmospheric and Solar-Terrestrial Physics, 180, 35
Hansteen, V., Guerreiro, N., De Pontieu, B., & Carlsson, M. 2015, ApJ, 811, 106
Harra, L. K., Sakao, T., Mandrini, C. H., et al. 2008, ApJ Letters, 676, L147
Hill, F. 2018, Space Weather, 16, 1488
Horbury, T. S., O'Brien, H., Carrasco Blazquez, I., et al. 2020, A&A, 642, A9
Jafarzadeh, S., Solanki, S. K., Feller, A., et al. 2013, A&A, 549, A116
Jafarzadeh, S., Solanki, S. K., Stangalini, M., et al. 2017, ApJ, Supplement, 229, 10
Judge, P. G., Habbal, S., & Landi, E. 2013, Solar Physics, 288, 467
Kasper, J. C., Abiad, R., Austin, G., et al. 2016, Space Science Reviews, 204, 131
Kasper, J. C., Klein, K. G., Lichko, E., et al. 2021, Physical Review Letters, 127, 255101
Kasper, J. C., Stevens, M. L., Korreck, K. E., et al. 2012, ApJ, 745, 162
Kasper, J. C., Stevens, M. L., Lazarus, A. J., Steinberg, J. T., & Ogilvie, K. W. 2007, ApJ, 660, 901
Klimchuk, J. A. 2006, Solar Physics, 234, 41
Klimchuk, J. A. 2015, Philosophical Transactions of the Royal Society of London Series A, 373, 20140256
Ko, Y.-K., Muglach, K., Wang, Y.-M., Young, P. R., & Lepri, S. T. 2014, ApJ, 787, 121
Krucker, S., Saint-Hilaire, P., Christe, S., et al. 2008, ApJ, 681, 644
Laming, J. M. 2015, Living Reviews in Solar Physics, 12, 2
Laming, J. M., Vourlidas, A., Korendyke, C., et al. 2019, ApJ, 879, 124
Lavraud, B., Ruffenach, A., Rouillard, A. P., et al. 2014, Journal of Geophysical Research (Space Physics), 119, 26
Lee, K.-S., Brooks, D. H., & Imada, S. 2015, ApJ, 809, 114
Linker, J. A., Caplan, R. M., Downs, C., et al. 2017, ApJ, 848, 70
Maksimovic, M., Bale, S. D., Chust, T., et al. 2020, A&A, 642, A12
Manchester, W. B., Gombosi, T. I., Roussev, I., et al. 2004, Journal of Geophysical Research (Space Physics), 109, A02107
Moise, E., Raymond, J., & Kuhn, J. R. 2010, ApJ, 722, 1411
Morton, R. J., Tiwari, A. K., Van Doorsselaere, T., & McLaughlin, J. A. 2021, ApJ, 923, 225
Müller, D., St. Cyr, O. C., Zouganelis, I., et al. 2020, A&A, 642, A1
Oran, R., Landi, E., van der Holst, B., Sokolov, I. V., & Gombosi, T. I. 2017, ApJ, 845, 98
Owen, C. J., Bruno, R., Livi, S., et al. 2020, A&A, 642, A16
Raouafi, N. E., Riley, P., Gibson, S., Fineschi, S., & Solanki, S. K. 2016, Frontiers in Astronomy and Space Sciences, 3, 20

Rast, M. P., Bello González, N., Bellot Rubio, L., et al. 2021, Solar Physics, 296, 70
Riley, P., Downs, C., Linker, J. A., et al. 2019, ApJ Letters, 874, L15
Rimmele, T. R., Warner, M., Keil, S. L., et al. 2020, Solar Physics, 295, 172
Rivera, Y. J., Landi, E., Lepri, S. T., & Gilbert, J. A. 2020, ApJ, 899, 11
Rochus, P., Auchère, F., Berghmans, D., et al. 2020, A&A, 642, A8
Rodríguez-Pacheco, J., Wimmer-Schweingruber, R. F., Mason, G. M., et al. 2020, A&A, 642, A7
Rouillard, A. P., Pinto, R. F., Vourlidas, A., et al. 2020, A&A, 642, A2
Ruffenach, A., Lavraud, B., Farrugia, C. J., et al. 2015, Journal of Geophysical Research (Space Physics), 120, 43
Schmelz, J. T., Reames, D. V., von Steiger, R., & Basu, S. 2012, ApJ, 755, 33
Shimojo, M. & Tsuneta, S. 2009, ApJ Letters, 706, L145
Solanki, S. K., del Toro Iniesta, J. C., Woch, J., et al. 2020, A&A, 642, A11
SPICE Consortium, Anderson, M., Appourchaux, T., et al. 2020, A&A, 642, A14
Stenborg, G., Howard, R. A., Vourlidas, A., & Gallagher, B. 2022, ApJ, 932, 75
Tomczyk, S., Card, G. L., Darnell, T., et al. 2008, Solar Physics, 247, 411
van der Holst, B., Manchester, W. B., I., Klein, K. G., & Kasper, J. C. 2019, ApJ Letters, 872, L18
Van Kooten, S. J. & Cranmer, S. R. 2017, ApJ, 850, 64
Vourlidas, A., Howard, R. A., Plunkett, S. P., et al. 2016, Space Science Reviews, 204, 83
Wang, S., Jenkins, J. M., Martinez Pillet, V., et al. 2020, ApJ, 892, 75
Winslow, R. M., Lugaz, N., Schwadron, N. A., et al. 2016, Journal of Geophysical Research (Space Physics), 121, 6092
Zhao, L., Landi, E., Lepri, S. T., et al. 2017, ApJ, 846, 135

The Era of Multi-Messenger Solar Physics
Proceedings IAU Symposium No. 372, 2023
G. Cauzzi & A. Tritschler, eds.
doi:10.1017/S1743921323001230

The Aditya-L1 mission of ISRO

Durgesh Tripathi[1], D. Chakrabarty[2], A. Nandi[3], B. Raghvendra Prasad[4], A. N. Ramaprakash[1], Nigar Shaji[5], K. Sankarasubramanian[4], R. Satheesh Thampi[6] and V. K. Yadav[6]

[1]Inter-University Centre for Astronomy and Astrophysics, Ganeshkhind, Pune 411007, India

[2]Physical Research Laboratory, Ahmedabad 380 009, India

[3]ISITE Campus, U R Rao ISRO Satellite Centre, Marathahalli, Bengaluru 560 037, India

[4]Indian Institute of Astrophysics, 2nd Block, Koramangala, Bengaluru 560 034, India

[5]Old Airport Road, Vimanapura PO Bengaluru 560017, India

[6]Space Physics Laboratory, VSSC, Thiruvananthapuram 695 022, India

Abstract. The Aditya-L1 is the first space-based solar observatory of the Indian Space Research Organization (ISRO). The spacecraft will carry seven payloads providing uninterrupted observations of the Sun from the first Lagrangian point. Aditya-L1 comprises four remote sensing instruments, *viz.* a coronagraph observing in visible and infrared, a full disk imager in Near Ultra-Violet (NUV), and two full-sun integrated spectrometers in soft X-ray and hard X-ray. In addition, there are three instruments for in-situ measurements, including a magnetometer, to study the magnetic field variations during energetic events. Aditya-L1 is truly a mission for multi-messenger solar astronomy from space that will provide comprehensive observations of the Sun across the electromagnetic spectrum and in-situ measurements in a broad range of energy, including magnetic field measurements at L1.

Keywords. Sun: abundances, Sun: activity, Sun: atmospheric motions, Sun: atmosphere, Sun: chromosphere, Sun: corona, Sun: coronal mass ejections (CMEs), Sun: faculae, plages, Sun: filaments, Sun: flares, Sun: fundamental parameters, Sun: general, Sun: granulation, Sun: infrared, Sun: magnetic fields, Sun: particle emission, Sun: photosphere, Sun: prominences, Sun: solar-terrestrial relations, Sun: solar wind, Sun: sunspots, Sun: UV radiation. Sun: X-rays

1. Introduction

The Sun is the life-giving star that has been studied for centuries. In the last 500 years or so, more detailed observations of the Sun's structure – both internal and atmosphere– have been studied. After discovering that the outer solar atmosphere was more than a million degrees hot in the 1940s, it was realized that to get a comprehensive understanding of the Sun, it is mandatory to go to space.

In the last several decades, several spacecraft have been flown by The National Aeronautics and Space Administration (NASA), the European Space Agency (ESA), and the Japan Aerospace Exploration Agency (JAXA). Some of the most important missions are Yohkoh (Acton et al. 1992), the Solar and Heliospheric Observatory (SoHO; Domingo et al. 1995), the Transition Region and Coronal Explorer (TRACE; Handy et al. 1999), the Solar Terrestrial Relations Observatory (STEREO; Socker et al. 2000), Hinode (Kosugi et al. 2007), the Solar Dynamics Observatory (SDO; Pesnell et al. 2012), the Interface Region Imaging Spectrometer (IRIS; De Pontieu et al. 2014) and the two most recently launched missions namely the Parker Solar Probe (PSP; Fox et al. 2016) and

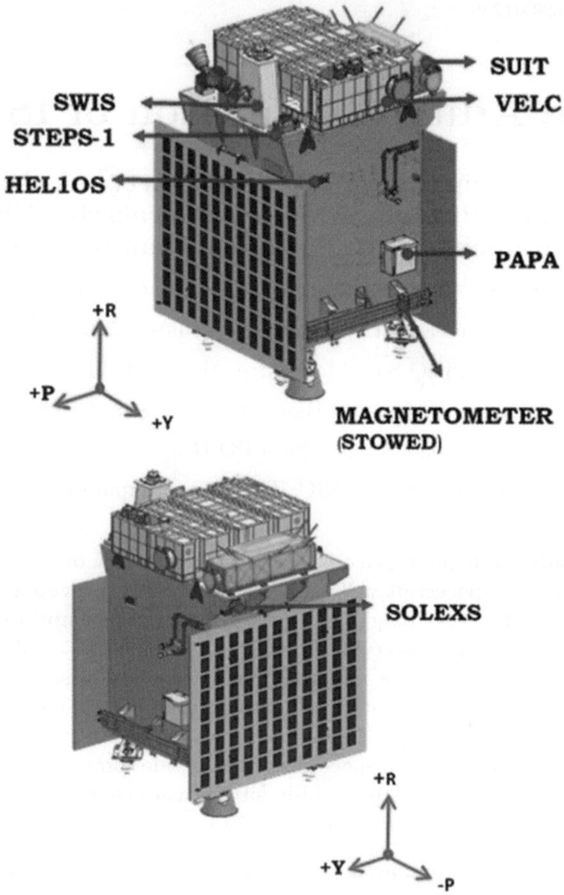

Figure 1. The model spacecraft indicating all the different payloads onboard Aditya-L1. Image courtesy of Indian Space Research Organization (ISRO) (`https://www.isro.gov.in/Aditya_L1.html`).

the Solar Orbiter (Müller et al. 2013, 2020). These, in addition to multiple ground-based observatories, have provided a wealth of information that has enhanced our understanding of the working of our star.

Although observations and modeling have allowed a great leap toward understanding the Sun, numerous unsolved problems remain. These problems relate to the coupling of the magnetized solar atmosphere, heating of the upper solar atmosphere, nature of solar wind, and dynamics of the inner heliosphere. Moreover, how and why the high energy radiation, particularly in the near ultraviolet, changes and affects the Earth's atmosphere is not fully comprehended.

One of the most challenging tasks related to coronal heating is measuring the coronal magnetic field. Currently, there is no direct measurement available. Such measurements will not only help understand the heating mechanism but also provide crucial information related to solar eruptions and possibly forecasting the direction of interplanetary magnetic field (IMF) associated with interplanetary coronal mass ejections (ICMEs). The other equally challenging task has been to determine the kinematic profiles of Coronal Mass Ejections (CMEs) during their early phases of evolution. Currently, regular kinematic profiles of CMEs are available using the coronagraphic observations made using the Large Angle and Spectrometric Coronagraph (C2 ad C3) on board SoHO. However, these are only available beyond 2.5 R⊙. However, observations have shown that 90% of the acceleration in CMEs occur below 2 R⊙ (Bein et al. 2011). Therefore, with the existing facilities, it is impossible to obtain a good handle on the early phase of the kinematics of CMEs. In turn, that becomes one of the primary sources of uncertainties in predicting the arrival time of CMEs. It has also been shown that a significant amount of activity, such as magnetic reconnections leading to the bifurcation of flux rope as well post CME reconnection, may occur well below 2.5 R⊙ (Gilbert et al. 2001; Tripathi et al. 2006, 2007).

CMEs are often associated with flares, which are the intense eruption of electromagnetic radiation. Flares are observed routinely in Soft X-rays using Geostationary Operational Environmental Satellite Network. However, it has been shown that they have counterparts across the electromagnetic spectrum (Benz 2017). While flares are specular in X-ray and have been used for monitoring the space weather, it has been shown that more than ∼70% of the flare energy is contained in other wavelengths such as visible and near-ultraviolet (Neidig 1989; Kretzschmar 2011). While this may have substantial implications on the physics of solar flares, it may have significant effects on the total and spectral irradiance of the Sun (Kretzschmar et al. 2010).

It is also known that the total solar irradiance varies slowly on decadal and longer time scales. The variation during the recent solar magnetic cycle has been about 0.1%†. However, it is known that solar irradiance shows a strong wavelength dependence and that up to 60% of the total irradiance variations are produced at wavelengths below 400 nm (Krivova et al. 2006). Of these, the wavelength band of 200–400 nm is of particular importance due to its direct effects on the chemistry of Ozone and Oxygen in the stratosphere of Earth. It thereby plays a crucial role in Sun-climate relations.

Using the observations taken by Solar Stellar Irradiance Comparison Experiment (SOLSTICE) and Spectral Irradiance Monitor (SIM), both on board Solar Radiation & Climate Experiment (SORCE), Haigh et al. (2010) showed that there is a considerable discrepancy between the solar spectral irradiance observed and modeled by Lean (2000). For the wavelength regions below 400 nm, the models under predicts the irradiance, while for the wavelengths larger than 400 nm, the irradiance is over-predicted. Such discrepancy may have a substantial impact on modeling the Sun-climate relations. Note that the observations are recorded using the Sun-as-a-star spectrum. These results suggest that some missing physics is crucial for accurately modeling the sun-climate relations.

The Aditya-L1 mission is the flagship mission of the Indian Space Research Organization (ISRO). It is ISRO's first solar observatory in space. The satellite will be launched in 2023 by PSLV-XL and inserted in a halo orbit around the First Lagrangian Point (L1), which is about 1.5 million kilometers from the Earth on the Sun-Earth line. The primary science goals of Aditya-L1 mission are:

- Diagnostics of the large and small scale structures in the solar corona
- Magnetic topology and field measurements in the solar corona
- Origin and dynamics of solar flares and coronal mass ejections (CMEs)
- Spectral energy distribution in solar flares
- Spatially resolved solar spectral irradiance
- Formation and dynamics of solar prominences and filaments
- Coronal abundances and first ionization potential (FIP) effects
- Solar wind composition and particle energy distribution
- Measurements of the magnitude and variability of the interplanetary magnetic field (IMF)

† Physikalisch-Meteorologisches Observatorium Davos and World Radiation Center (PMOD/WRC)

To address the above-mentioned questions, the spacecraft will carry seven payloads performing remote sensing from Hard X-ray to infrared and in situ measurements in a broad energy range, including interplanetary magnetic field measurements. The seven payloads are:

- The Solar Ultraviolet Imaging Telescope (SUIT),
- The Visible Emission Line Coronagraph (VELC),
- The Solar Low Energy Spectrometer (SoLEXS),
- The High Energy L1 Orbiting X-ray Spectrometer (HEL1OS),
- The Aditya Solar Wind Particle EXperiment (ASPEX),
- The Plasma Analyser Package for Aditya (PAPA), and
- The Magnetometer (MAG)

In the rest of the paper, we shall discuss and describe the uniqueness and salient features of different payloads and the science questions they aim to address.

2. Payloads

The Aditya-L1 spacecraft carries seven payloads – four for remote sensing and three for in-situ measurements. Below we describe each of the payloads.

2.1. *The Solar Ultraviolet Imaging Telescope (SUIT)*

The Solar Ultraviolet Imaging Telescope (SUIT; Tripathi et al. 2017; Ghosh et al. 2016, 2022) is one of the four remote sensing payloads on board Aditya-L1. It is a combined medium and narrow band filter imager that will observe the Sun in the Near UltraViolet (NUV) wavelength band 200–400 nm. With this wavelength range, SUIT aims to cover the lower and middle atmospheric layer of the Sun, i.e., the photosphere and the chromosphere. These two layers are the most crucial for addressing questions related to the dynamic coupling of the magnetized solar atmosphere and various fundamental processes occurring in the partially ionized plasma. Moreover, solar radiation in this wavelength range is fundamental to studying Sun-Climate relations. Therefore, SUIT will make unique measurements of spatially resolved Solar Spectral Irradiance (SSI) in the NUV. Such measurements are crucial for comprehending the chemistry of Ozone and Oxygen in the stratosphere of the Earth and, therefore, central to the Sun-climate relationship.

The primary science questions that SUIT will address are:

- the magnetic coupling between the lower and the middle solar atmosphere,
- the spatially resolved SSI in the NUV (200–400 nm),
- the energetics of solar flares in the photosphere and chromosphere,
- the dynamics of eruptive and non-eruptive prominences and filaments and
- to provide the context to the VELC.

SUIT is a Ritchey-Chrétien telescope with folded optics carrying a primary mirror with an aperture size of 140 mm and a secondary mirror. It has a set of 11 different science filters centered around different wavelengths between 200–400 nm (see Table. 1). The images are recorded on a CCD that is kept at the focus. The CCD is UV-thinned and back-illuminated. It has a full well capacity of 190000 e^- and a dynamic range of ~12000. SUIT will provide partial disk and full disk images of the Sun with a pixel size of 0.7" and temporal resolution that can range between 4 to 40 seconds, depending on the modes of operation, slicing through the solar atmosphere using various filters. The telescope will produce images with a signal-to-noise (SNR) ratio of 100 in the darkest region of the Sun and a contrast ratio of 10:1 under ambient conditions.

Table 1. Filter abbreviations, central wavelength, and bandpass of the SUIT science filters.

Index	Filter name	Central Wavelength (nm)	Bandpass (nm)
1	NB1	214.0	11.0
2	NB2	276.7	0.4
3	NB3	279.6	0.4
4	NB4	280.3	0.4
5	NB5	283.2	0.4
6	NB6	300.0	1.0
7	NB7	388.0	1.0
8	NB8	396.85	0.1
9	BB1	220.0	40.0
10	BB2	277.0	58.0
11	BB3	340.0	40.0

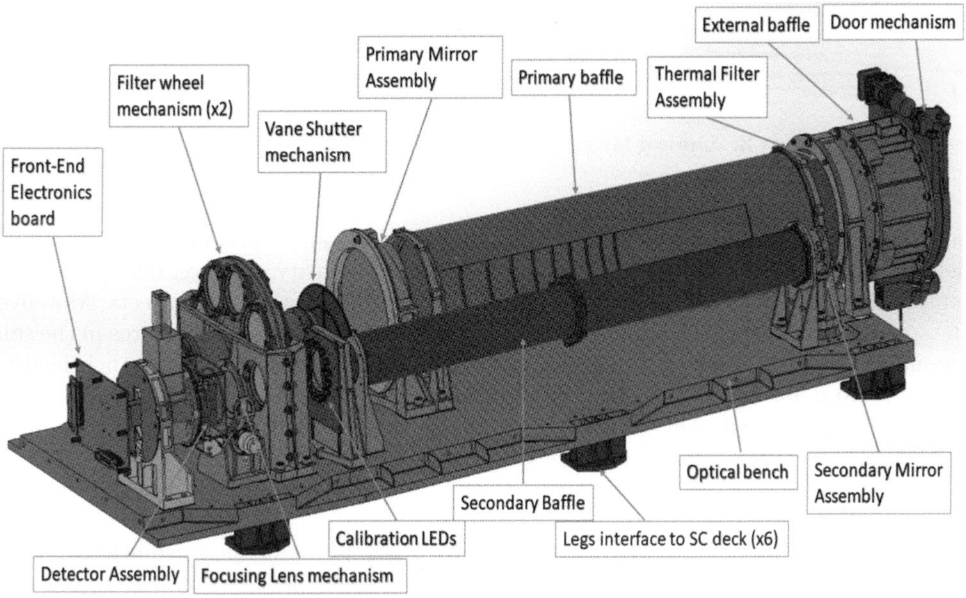

Figure 2. Internal assembly view of the Solar Ultraviolet Imaging Telescope (SUIT).

SUIT has an inbuilt onboard intelligence for tracking regions of interest and detection, localization, and automatic exposure control for flare observations. See Tripathi et al. (2017); Ghosh et al. (2016, 2022) for further details on the instrument.

2.2. *Visible Emission Line Coronagraph (VELC)*

The Visible Emission Line Coronagraph (VELC; Raghvendra Prasad et al. 2017) is an internally occulted coronagraph on board Aditya-L1. It can perform simultaneous imaging, spectroscopy, and spectro-polarimetric observations in the lower corona. VELC will image the corona in white light from 1.05–3.0 $R_\odot$ with a plate scale of 2.5 arcsec/pixel. Spectroscopic and spectro-polarimetric observations will be performed using a multi-slit spectrograph in three emission lines formed at 530.3, 789.2, and 1074.7 nm, with a spectral resolution of 28, 31, and 202 mÅ/pixel. The imaging plate scale for 530.3 nm and 789.2 nm is 1.25 arcsec/pixel and that for 1074.7 nm is 4.8 arcsec/pixel. The field of view (FOV) for the spectroscopic and spectro-polarimetric channels will be from 1.05–1.5 $R_\odot$. While the imaging observations in white-light will provide crucial information

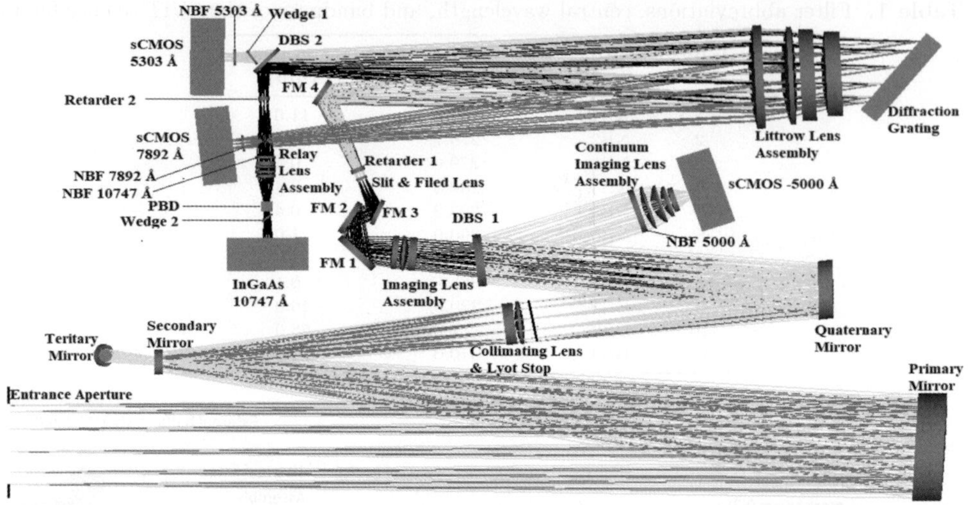

Figure 3. Optical layout of the Visible Emission Line Coronagraph.

on the early evolution of CMEs and its kinematics, the spectroscopic observations will help us measure the plasma parameters such as line intensity, electron density, electron temperature and thermal and non-thermal spectral line widths in the ejects. Moreover, the spectroscopic measurements of plasma parameters on quiescent structures in the solar corona will provide crucial information for modeling coronal structures and their heating. The spectro-polarimetric measurements in Fe XIII 1074.7 nm infrared line will provide the topology of the coronal magnetic field for the first time from space.

The primary science questions that VELC will address are:

• What is the most dominant mechanism for coronal heating? What role do MHD waves play?

• What is the kinematics of CMEs during the early phase of their evolution?

• To what degree do coronal inhomogeneities affect the heating and acceleration processes?

• What is the magnetic field configuration in the lower corona?

VELC has an entrance aperture of 148 mm and an off-axis parabolic primary mirror of 200 mm diameter (see Fig. 3). The light from the primary mirror is then reflected to the secondary mirror, that act as an internal occulter. It allows the disk light to pass through and reflects only the coronal light that is further used for imaging and spectroscopy. The disk light is reflected out into space using another mirror. For imaging and spectroscopy in visible, VELC uses three sCMOS detectors. However, for spectro-polarimetry using Fe XIII in infrared, VELC uses InGaAs detector system. See Raghvendra Prasad et al. (2017); Venkata et al. (2022); Singh et al. (2019) for further details on the instrument. Like SUIT, VELC also developed an onboard algorithm for automatic CME detection Patel et al. (2018).

2.3. *X-Ray Observations of the Sun with Aditya-L1*

The Aditya-L1 mission has two full-Sun integrated X-ray spectrometers (Sankarasubramanian et al. 2017), *viz.* the Solar Low Energy X-ray Spectrometer (SoLEXS) and High Energy L1 Orbiting X-ray Spectrometer (HEL1OS) with their primary goal of studying solar flares.

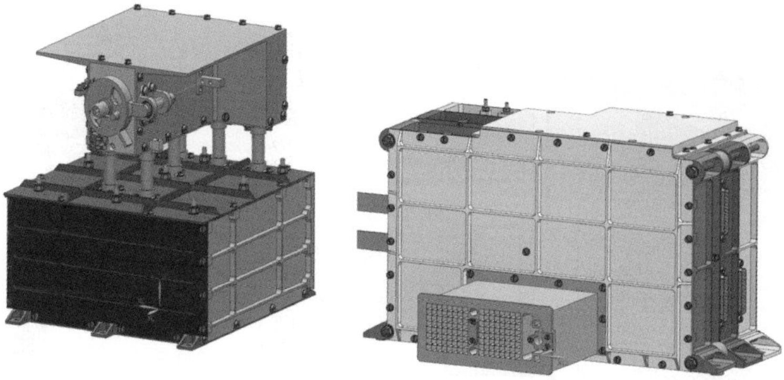

Figure 4. Concept design of SoLEXS (Left panel) and HEL1OS (right panel).

SoLEXS (see left panel of Fig. 4) covers the energy band from 1 to 30 keV with a spectral resolution of <250 eV at 6 keV and primarily covers the thermal energy part of the flares. SoLEXS carries two identical detectors but with different aperture sizes to observe smaller as well as larger flares. The detector with the larger aperture will be sensitive to smaller flares, whereas that with the smaller aperture will be sensitive to larger flares. It will have onboard processing to help provide spectra at a cadence of 1s during flares. Moreover, it will have onboard intelligence to detect the start of flares. This input is also provided to the SUIT payload for their algorithm to localize and start the flare mode operation.

HEL1OS (see right panel of Fig. 4) covers the energy band from 10 to 150 keV, with the prime aim of studying the impulsive phase of the flares, where most of the emission is shown to be non-thermal. Therefore, it will provide unique observations to help identify the non-thermal energy released during flares. HEL1OS will carry two different types of detectors, *CZT and CdTe.* There are two CZT detectors that will cover the energy range of 20 to 150 keV. The CdTe detector is sensitive from 10 to 40 keV and will be used for detailed spectroscopic studies. For further details of SoLEXS and HEL1OS, see Sankarasubramanian et al. (2017).

The two spectrometers together will cover a broad range of energy, i.e., from 1 keV to 150 keV, with excellent spectral resolution. The primary science objectives of the two spectrometers together are:

- explosive energy release, particle acceleration, and transport using high-resolution spectral and imaging information.
- trigger mechanism of flares/eruptive processes.
- cut-off energy between thermal and non-thermal emission as a function of flare evolution.
- SXR, HXR and CME association

2.4. *In-situ measurements at L1 with Aditya-L1*

Among the seven payloads, three are *in situ* instruments, namely the Aditya Solar Wind Particle Experiment (ASPEX), the Plasma Analyser Package for Aditya (PAPA), and Magnetometer (MAG). While ASPEX and PAPA are particle detectors, the MAG instrument is flux gate magnetometers (Janardhan et al. 2017).

ASPEX (see Fig. 5) contains two ion spectrometers (Janardhan et al. 2017; Goyal et al. 2018; Mayank et al. 2022), *viz.* the Solar Wind Ion Spectrometer (SWIS) and the Supra Thermal and Energetic Particle Spectrometer (STEPS). While SWIS covers the low energy particles in the range 0.1–20 keV, STEPS aims to detect high energy particles

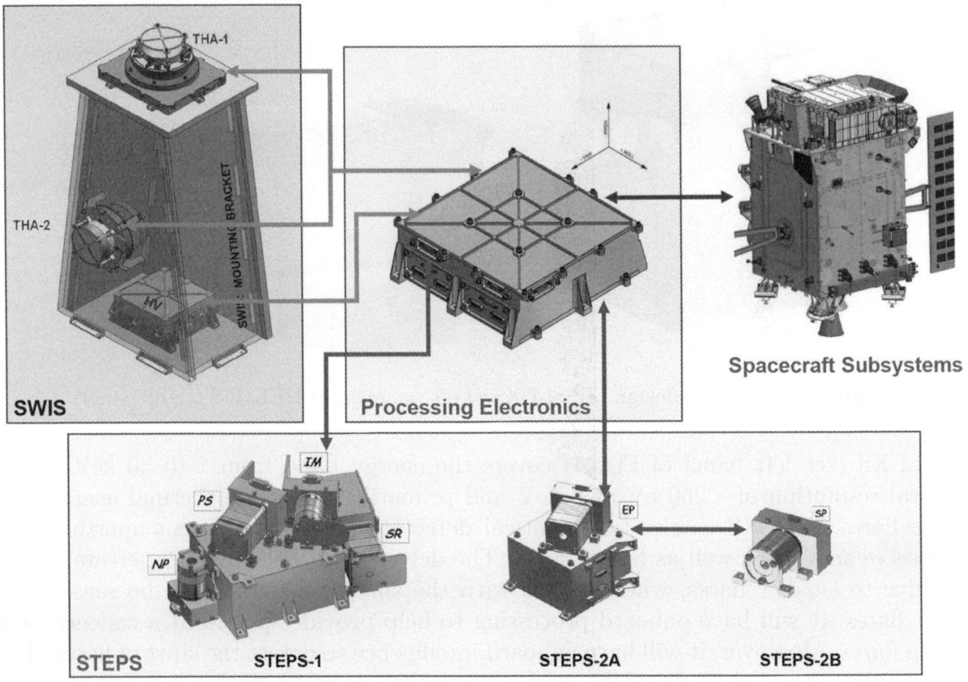

Figure 5. ASPEX payload and its various subsystems. THA, HV stand for Top Hat Analyzer and High Voltage unit respectively. Note, SWIS consists of THA-1 and THA-2 packages and STEPS consists of STEPS-1 (Sun Radial or SR, Intermediate or IM, Parker Spiral or PS, North Pointing or NP) and STEPS-2 (Earth Pointing or EP and South Pointing or SP) packages.

in the energy range 20 keV/nucleon–5 MeV/nucleon. The primary science objectives of ASPEX are to understand better the origin, acceleration, inter-relationship and anisotropy of ions (primarily protons and alphas) in the background solar wind during the passage of transient events such as CMEs and co-rotating interaction regions (CIRs). By measuring (protons and alphas separated in some directions and integrated in other directions) these ions, in the slow and fast solar wind (360 deg along and across the ecliptic) as well as in the suprathermal and solar energetic particle (SEP) domains in multiple directions (Sun Radial, Intermediate, Parker Spiral, North Pointing, South Pointing and Earth Pointing), ASPEX will be able to investigate the characteristics of CMEs, SIRs (or CIRs) and energetic ions. Further, based on the direction-resolved observations of these ions, ASPEX may pave the way for the efficient prediction capability for the arrival of these transients at the L1 point.

The Plasma Analyzer Package for Aditya (PAPA) aims at studying the composition of solar wind and its energy distribution (in the range from 0.01–3 keV for electrons and 0.01–25 keV for ions) continuously throughout the mission's lifetime. PAPA contains two sensors (see Fig. 6); Solar Wind Electron Energy Probe (SWEEP) to measure the solar wind electron flux and Solar Wind Ion Composition AnalyseR (SWICAR) to measure the ion flux and composition as a function of direction and energy. SWEEP will measure only electron parameters whereas SWICAR has two modes of operation – ion mode where ion parameters are measured and electron mode where electron parameters are measured. These two modes in SWICAR are mutually exclusive. Data from PAPA would provide detailed knowledge of the solar wind condition with high time resolution. SWICAR would also provide: (1) the elemental composition of solar wind low energy ions, and (2) the differential energy spectra, abundances of dominant ion species in the mass range 1-60

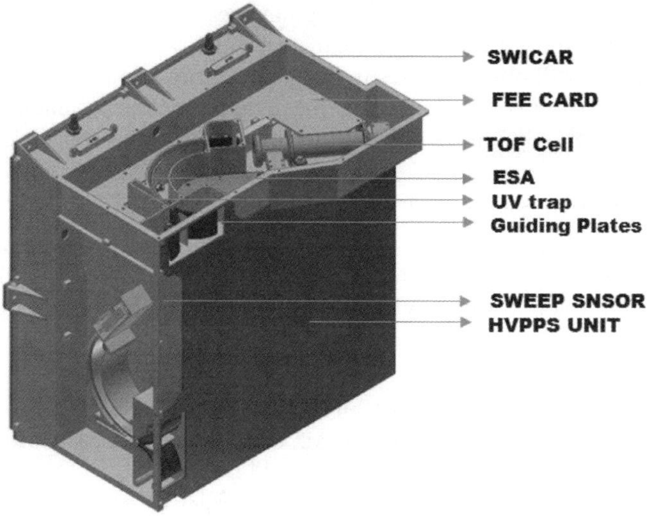

Figure 6. Schematics of the PAPA payload. Different components are marked.

amu with energies between 0.01–25 keV/q. From the differential energy spectra, the 'key parameters' such as bulk speed, density and kinetic temperature of the dominant solar wind ion species could be regularly derived; inferences could be made on the coronal temperatures; plasma sources of suprathermal ion populations could be identified, and the nature and dynamics of solar wind plasma could be quantified.

The MAG instrument package consists of two tri-axial flux gate magnetometers (FGMs). The two FGMs will be mounted on a 6-meter-long boom. The two FGMs are used because the spacecraft is not magnetically clean. Therefore, the presence of two identical FGMs is used to obtain the differential interplanetary magnetic field (IMF) by nullifying the effect of the magnetic field from the spacecraft. The primary science objective of MAG is to measure and monitor the magnitude and direction of the IMF locally at the L1 point. The magnitude and direction of IMF associated with CMEs are two crucial parameters for space weather forecasting. The other important science objective of the MAG payload is to observe the CMEs from the Sun coming towards the Earth at the L1 point (Yadav et al. 2018).

Apart from the above primary objectives, the secondary objectives of this magnetic field experiment are (1) Near Earth Space Weather studies thereby combining the in-situ magnetic field measurements onboard Aditya-L1 spacecraft with ground based measurements of geomagnetic field; and (2) the detection of plasma wave signatures at the L1 point with the magnetic field and solar wind particle velocity measurements. These plasma (possibly Alfvén) wave signatures are detected in the magnetic field and solar wind observations onboard ACE and WIND (Yadav 2020).

The measurements from ASPEX and PAPA combined with those from MAG will provide crucial insights towards understanding the dynamics of the inner heliosphere and physical processes occurring therein and will help us towards better forecasting of space weather events.

3. Summary

Aditya-L1 is truly a satellite for multi-messenger solar physics. With the suite of instruments on board, Aditya-L1 will perform remote sensing covering a broad range of the electromagnetic spectrum and in situ measurements in a wide energy range from the same platform. For the first time, it will provide regular measurements of the coronal magnetic field from space and spatially resolved solar spectral irradiance in the near ultraviolet. Moreover, Aditya-L1 will observe flares in both soft X-ray and hard X-ray from the same platform. Similarly, the two particle instrument, along with the magnetometer, will be highly beneficial for the dynamics of the inner heliosphere and space weather studies.

Acknowledgements

The generous travel support from the IAU is gratefully acknowledged. Aditya-L1 is a mission of the Indian Space Research Organization (ISRO). We acknowledge ISRO's all round support in developing the payloads and proving this mission of opportunity.

References

Acton, L., Tsuneta, S., Ogawara, Y., et al. 1992, Science, 258, 618
Bein, B. M., Berkebile-Stoiser, S., Veronig, A. M., et al. 2011, ApJ, 738, 191
Benz, A. O. 2017, Living Reviews in Solar Physics, 14, 2
De Pontieu, B., Title, A. M., Lemen, J. R., et al. 2014, Solar Physics, 289, 2733
Domingo, V., Fleck, B., & Poland, A. I. 1995, Space Science Reviews, 72, 81
Fox, N. J., Velli, M. C., Bale, S. D., et al. 2016, Space Science Reviews, 204, 7
Ghosh, A., Chatterjee, S., Khan, A. R., et al. 2016, in Society of Photo-Optical Instrumentation Engineers (SPIE) Conference Series, Vol. 9905, Space Telescopes and Instrumentation 2016: Ultraviolet to Gamma Ray, ed. J.-W. A. den Herder, T. Takahashi, & M. Bautz, 990503
Ghosh, A., Kesharwani, R., Khan, A. R., et al. 2022, in Society of Photo-Optical Instrumentation Engineers (SPIE) Conference Series, Vol. 12181, Society of Photo-Optical Instrumentation Engineers (SPIE) Conference Series, ed. J.-W. A. den Herder, S. Nikzad, & K. Nakazawa, 121813O
Gilbert, H. R., Holzer, T. E., & Burkepile, J. T. 2001, ApJ, 549, 1221
Goyal, S. K., Kumar, P., Janardhan, P., et al. 2018, Planetary and Space Science, 163, 42
Haigh, J. D., Winning, A. R., Toumi, R., & Harder, J. W. 2010, Nature, 467, 696
Handy, B. N., Acton, L. W., Kankelborg, C. C., et al. 1999, Solar Physics, 187, 229
Janardhan, P., Vadawale, S., Bapat, B., et al. 2017, Current Science, 113, 620
Kosugi, T., Matsuzaki, K., Sakao, T., et al. 2007, Solar Physics, 243, 3
Kretzschmar, M. 2011, A&A, 530, A84
Kretzschmar, M., de Wit, T. D., Schmutz, W., et al. 2010, Nature Physics, 6, 690
Krivova, N. A., Solanki, S. K., & Floyd, L. 2006, A&A, 452, 631
Lean, J. 2000, GRL, 27, 2425
Mayank, P., Vaidya, B., & Chakrabarty, D. 2022, ApJS, 262, 23
Müller, D., Marsden, R. G., St. Cyr, O. C., Gilbert, H. R., & Solar Orbiter Team. 2013, Solar Physics, 285, 25
Müller, D., St. Cyr, O. C., Zouganelis, I., et al. 2020, A&A, 642, A1
Neidig, D. F. 1989, Solar Physics, 121, 261
Patel, R., Amareswari, K., Pant, V., et al. 2018, Solar Physics, 293, 103
Pesnell, W. D., Thompson, B. J., & Chamberlin, P. C. 2012, Solar Physics, 275, 3
Raghvendra Prasad, B., Banerjee, D., Singh, J., et al. 2017, Current Science, 113, 613
Sankarasubramanian, K., Sudhakar, M., Nandi, A., et al. 2017, Current Science, 113, 625
Singh, J., Prasad, B. R., Venkata, S., & Kumar, A. 2019, Advances in Space Research, 64, 1455
Socker, D. G., Howard, R. A., Korendyke, C. M., Simnett, G. M., & Webb, D. F. 2000, in Society of Photo-Optical Instrumentation Engineers (SPIE) Conference Series, Vol. 4139, Instrumentation for UV/EUV Astronomy and Solar Missions, ed. S. Fineschi, C. M. Korendyke, O. H. Siegmund, & B. E. Woodgate, 284–293
Tripathi, D., Ramaprakash, A. N., Khan, A. R., et al. 2017, Current Science, 113, 616
Tripathi, D., Solanki, S. K., Mason, H. E., & Webb, D. F. 2007, A&A, 472, 633

Tripathi, D., Solanki, S. K., Schwenn, R., et al. 2006, A&A, 449, 369
Venkata, S., Prasad B, R., & Singh, J. 2022, Experimental Astronomy, 53, 71
Yadav, V. K. 2020, IETE Technical Review, 37, 3
Yadav, V. K., Srivastava, N., Ghosh, S. S., Srikar, P. T., & Subhalakshmi, K. 2018, Advances in Space Research, 61, 749

The Era of Multi-Messenger Solar Physics
Proceedings IAU Symposium No. 372, 2023
G. Cauzzi & A. Tritschler, eds.
doi:10.1017/S1743921322004586

Solar Observations with ALMA: A New Frontier

T. S. Bastian

National Radio Astronomy Observatory
520 Edgemont Rd., Charlottesville, VA 22903 USA

Abstract. Solar observations with the Atacama Large Millimeter-Submillimeter Array (ALMA) became available to the community in late-2016. For the first time, high angular resolution (sub-arcsec) and high-time-resolution (1 s) observations of the Sun became possible at millimeter wavelengths, providing observations of the solar chromosphere that are uniquely complementary to those in O/IR and UV wavelengths. Here, an overview of current ALMA capabilities is provided, selected recent results of ALMA observations of the Sun are highlighted, and future capabilities are outlined.

Keywords. Sun: radio radiation, Sun: chromosphere; instrumentation: interferometers

1. Introduction

The Atacama Large Millimeter-submillimeter Array (ALMA) is a general purpose telescope designed to open a previously under-explored electromagnetic window. It is an interferometer designed and optimized to provide high resolution imaging over a broad range of wavelengths, currently from 0.32-3 mm (92-950 GHz). As such, ALMA addresses an extremely broad science program spanning solar system science; planet, star, and galaxy formation and evolution; astrochemistry; and cosmology (Wootten & Thompson 2009). ALMA is located in the high desert plains of the Atacama desert in northern Chile at an elevation of 5000 m (Fig. 1). It comprises 66 antennas: 50×12 m antennas that are used in a reconfigurable array for interferometry (INT), referred to as the 12-m array; 12×7 m antennas that are fixed in position, referred to as the Atacama Compact Array (ACA), also used for interferometry; and four total power (TP) antennas that can be used for low-resolution single dish total power mapping.

ALMA is a synthesis imaging telescope. That is, it measures the Fourier transform of the intensity distribution within its field of view. A single pair of antennas measures a single Fourier component at a particular spatial frequency determined by the antenna separation and orientation (the antenna baseline) and the wavelength being observed. Measurements are made in *uv* plane, *u* and *v* referring to the spatial frequency coordinates of a given antenna baseline. Fourier inversion of the *uv* data yields a "dirty map", so called because it represents the "true" intensity distribution convolved with the instrument point spread function (PSF) or "dirty beam". Deconvolution techniques are used to remove the effects of the PSF and to estimate the true distribution of intensity within the field of view. Further details regarding synthesis imaging may be found in, e.g., Taylor et al. (1999). ALMA observations in 8 frequency bands are available for general use but, as we discuss below, only 4 of these bands are available for solar observations. A given frequency band is divided into 4 spectral windows, each 2 GHz in bandwidth. Each spectral window, in turn, is typically divided into 128 spectral channels to enable the

Figure 1. The Atacama Large Millimeter-submillimeter Array. Credit: ALMA (ESO/NAOJ/NRAO).

removal of unwanted spectral line signals (e.g., terrestrial O_2 or H_2O) and to correct for the frequency bandpass response. The data may then be averaged in frequency to produce pseudo-continuum datasets in each of the spectral windows. Currently, solar observations are made in the two orthogonal senses of linear polarization but polarimetry is not yet supported.

Solar observations with ALMA pose many challenges. The Sun fills the ALMA field of view with bright emission on a wide range of angular scales. Moreover, the emission is time variable due to the Sun's apparent motion, its rotation, and intrinsic variations (e.g., oscillations, transient outbursts). Shimojo et al. (2017) detail the technical changes to the system needed to allow ALMA to make observations of the Sun without saturating the receivers or other elements of the electronics chain, and to calibrate the INT data. Similarly, White et al. (2017) describe how TP mapping observations of the Sun with ALMA are performed. Bastian et al. (2022) discusses data analysis challenges encountered by users working with ALMA solar observations. We therefore do not discuss technical, calibration, or data analysis issues here. Instead, this paper begins with a brief discussion of why ALMA is relevant to solar physics in §2. Current observing capabilities are then summarized in §3 and a number of recent science results are highlighted in §4. Examples of new observing capabilities that are under active development, or soon will be, are discussed in §5. We conclude in §6.

2. Why ALMA?

At mm-λ the source opacity in the solar atmosphere is dominated by H and He free-free absorption. At sub-mm-λ H^- opacity – that is, collisions of electrons on neutral H – also becomes significant. In both cases the emission occurs under conditions of local thermodynamic equilibrium (LTE) and the source function is therefore Planckian. Furthermore, the Rayleigh-Jeans approximation is valid and so the observed intensity (or brightness temperature) is linearly proportional to the kinetic temperature of the emitting plasma for optically thick emission. The radiative transfer is consequently straightforward. Note, however, that while the source function is in LTE the degree of H and He ionization can depart significantly from ionization equilibrium and so the opacity may depart from its LTE value.

Solar mm/sub-mm-λ emissions largely originate in the chromosphere, with the shortest wavelengths emitted from lower in the chromosphere and longer wavelengths emitted higher in the chromosphere. The chromosphere is complex and poorly understood. It is the layer in which the onset of non-radiative heating occurs; where waves, shocks, and

Table 1. ALMA Solar Observing Bands and Spectral Windows.

Freq. Band Designation	ν_{LO} (GHz)	λ_{LO} (mm)	spw 0 (GHz)	spw 1 (GHz)	spw 2 (GHz)	spw 3 (GHz)
3	100	3	92-94	94-96	104-106	106-108
5	198	1.51	190-192	192-194	202-204	204-206
6	239	1.25	229-231	231-233	245-247	247-249
7	346.6	0.86	338.6-340.6	340.6-342.6	350.6-352.6	352.6-345.6

mass motions drive complex dynamics; where there is a transition from the dominance of gas pressure to magnetic pressure; and where there is a transition from a low degree of ionization to a high degree of ionization. Outstanding problems include the structure and dynamics of the chromosphere, its energy and mass budget, and the transfer of energy and mass between the photosphere, chromosphere, and the corona. Observations at mm/sub-mm-λ provide a powerful complement to those in optical (O) and ultraviolet (UV) wavelengths, which generally form under conditions of non-LTE and for which the radiative transfer is complicated. Any model of the chromosphere must simultaneously account for both mm/sub-mm-λ continuum emission and the O/UV line emission.

The potential of mm/sub-mm-λ emission as a probe of the solar atmosphere has been recognized for many years — continuum observations date back to the 1960s (see, for example, the compilation of brightness temperature measurements by Linsky 1973). However, these and more modern observations (e.g., Bastian et al. (1993)) have been severely hampered by a lack of angular resolution since most observations were performed by single dish telescopes with apertures of no more than 10-15 m. Hence, while single dish measurements provided useful constraints to semi-empirical models of the solar atmosphere – e.g., Vernazza et al. (1981) – mm/sub-mm-λ/ wavelengths remained severely under-utilized.

With the advent of ALMA the situation has changed. Arcsecond imaging on time scales as short as 1 s is now possible in four frequency bands spanning wavelengths of 0.85 – 3 mm. ALMA observations are proving to be particularly interesting when used in concert with observations at optical, UV, and EUV wavelength using *Hinode* and ground based observatories, the Interface Region Imaging Spectrometer (IRIS), and the Solar Dynamics Observatory (SDO), respectively. We now turn to the ALMA solar observing capabilities currently available.

3. Current Solar Observing Capabilities

Current ALMA solar observing modes offer a powerful range of capabilities for quiet Sun science. The focus to date has been in providing capabilities to the community that allow high-fidelity continuum imaging over a broad range of frequencies with arcsecond (0.6" – 0.9") angular resolution and with a time resolution as high as 1 s. The possibility of observing solar flares and other new observing modes is discussed in §5.

3.1. *Frequency Bands*

Four frequency bands are currently offered to the solar community, as summarized in Table 1. Initially, in 2016 (ALMA Cycle 4), only continuum observations in Band 3 (3 mm) and in Band 6 (1.25 mm) were available. Band 7 (0.86 mm) was made available to the community in 2019 (Cycle 7) and Band 5 (1.51 mm) was to be made available in 2020 (Cycle 8). However, the pandemic led to a suspension of telescope operations in 2020-21 that curtailed Cycle 7 and effectively cancelled Cycle 8. While ALMA resumed science operations in October 2021, the number of solar programs that have been executed has been limited. Hence, observations in Bands 5 and 7 have been very limited to date.

Table 2. ALMA Angular Resolution and Maximum Recoverable Scale.

	Freq. Band	**3**	**5**	**6**	**7**
Configuration	ν **(GHz)/**λ **(mm)**	**100/3**	**198/1.51**	**239/1.25**	**346.6/0.86**
ACA	θ_{res}(arcsec)	11.6	6.77	5.45	3.63
	θ_{MRS} (arcsec)	66.0	36.1	29.0	19.3
C-1	θ_{res}(arcsec)	3.38	1.83	1.47	0.98
	θ_{MRS} (arcsec)	28.5	15.4	12.4	8.25
C-2	θ_{res}(arcsec)	2.30	1.24	1.00	0.67
	θ_{MRS} (arcsec)	22.6	12.2	9.81	6.54
C-3	θ_{res}(arcsec)	1.42	0.77	0.62	
	θ_{MRS} (arcsec)	16.2	8.73	7.02	
C-4	θ_{res}(arcsec)	0.92			
	θ_{MRS} (arcsec)	11.2			

3.2. *Array Configurations*

The Sun fills ALMA's field of view for all pointings on the solar disk with complex emission over a wide range of angular scales. As an interferometer, ALMA can only sample the angular scales sampled by the independent antenna baselines. The angular resolution of the instrument is determined by the maximum antennas baselines, whereas the maximum angular scales measured by the array are determined by the minimum antenna baselines. As a means of addressing a wide range of astronomical imaging problems, the 12 m array is reconfigurable, cycling over ten array configurations. However, of these, only the four most compact 12 m array configurations are available for solar observing. There are two reasons for this limitation. From an imaging perspective, the larger the array configuration and the higher the observing frequency, the sparser the sampling of the *uv* domain, thereby degrading imaging fidelity. From a calibration perspective, limits are imposed by precipitable water vapor in the atmosphere overlying the telescope by introducing phase variations in the wave front incident on each antenna. These variations can propagate into the visibility data as a loss of coherence with increasing baseline length unless corrected. ALMA antennas are outfitted with water vapor radiometers that measure brightness temperature variations of the sky over each antenna, a proxy for phase variations. Unfortunately, the radiometers saturate when pointing at the Sun and so such measurements are not available for solar observations. The impact of the phase variations increases with antenna baselines and so more compact configurations are favored, namely configurations C-1, C-2, C-3, and C-4. The maximum baselines in each configuration are 161, 314, 500, and 783 m, respectively. Table 2 gives the angular resolution and the "maximum angular scale" recovered by each of the array configurations currently available for solar observations. Note that Band 7 observations are limited to the two most compact arrays whereas Band 3 can use all four of the most compact configurations. Note, too, the parameters for the ACA are provided. The fact that θ_{MRS} decreases with increasing configuration number is partly mitigated by using both the fixed ACA 7 m antennas and the 12-m array as a heterogeneous array. Hence, "standard" solar observations with ALMA use the heterogeneous array for all INT observations.

3.3. *Total Power Mapping*

All interferometers behave as high pass spatial filters because they only measure spatial frequencies corresponding to baselines greater than some minimum. Hence, information on the largest angular scales is missing, including the total flux. This is particularly important for the Sun because most of the flux is on the largest spatial scales. Accurate photometry requires that it be recovered. The problem of missing information on the largest angular scales is partially mitigated by the use of the heterogeneous array, but

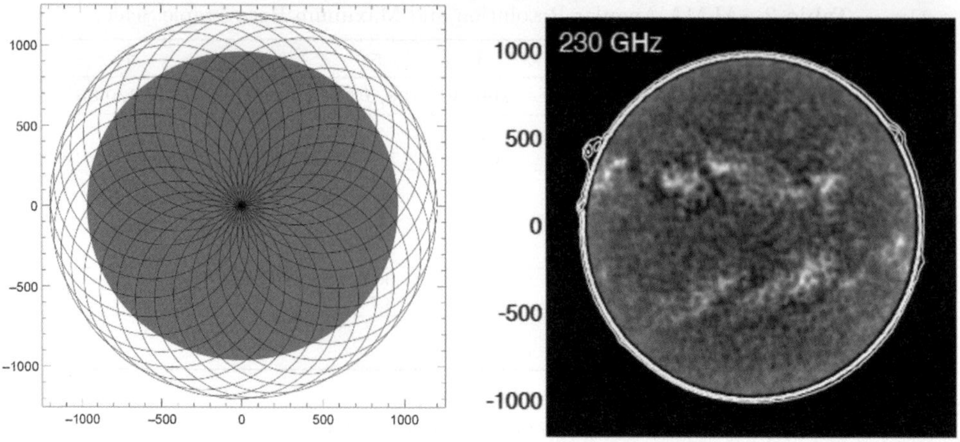

Figure 2. Left: schematic representation of the double-circle fast-scan mapping pattern used to produce full disk TP maps with ALMA. The grey disk represents the solar disk. Right: An example of a Band 6 TP map (from White et al. 2017).

information is still missing. The missing information – the "hole" in the *uv* plane – is addressed with TP maps. As discussed by White et al. (2017), fast-scan total power mapping can be used to produce full disk maps of the Sun with the resolution of a 12 m TP antenna ($\theta_{FOV} = 19.4" \lambda_{mm}$). Such maps can be used to, in effect, fill in the shortest spatial frequencies and the total power flux, thereby enabling maps to be recovered that contain flux all angular scales, from the field of view to the resolution defined by the maximum baseline. Figure 2 shows a schematic of the fast scanning pattern used to perform low-resolution TP mapping. An example of a Band 6 TP map is also shown.

Recently, fast regional mapping (FRM) was commissioned and made available to the solar community for Cycle 9. The FRM mode allows users to perform rapid (10s of sec) mapping of a users-specified region of limited size in a continuous fashion. For example, a circular region 3 arcmin in diameter – an active region – could be mapped at low resolution every 20 s. This mode will be particularly important for time variable phenomena.

3.4. *Single Field Imaging and Mosaicking*

The instantaneous field of view of ALMA is largely determined by the 12 m antennas (θ_{FOV}; for the heterogeneous array, the effective FOV is $\approx 6\%$ broader). For certain science objectives, this may be sufficient. For example, the minimum cadence at which snapshot imaging can be performed is 1 s. This may be attractive for observations of certain variable phenomena. For many science objectives, however, the limited FOV may be inadequate and observers must resort to "mosaicking" techniques. As the name implies, mosaicking involves sampling a user-specified pattern of independent telescope pointings, spatially separated by no more than that required for Nyquist sampling. These pointings are then assembled in post-processing into an image that can be significantly larger than the FOV of a single antenna. At present, mosaics are limited to 150 pointings which yields a map size of approximately $33" \times 33" \lambda^2_{mm}$; i.e., $5' \times 5'$ at a wavelength of 3 mm. Other mapping aspect ratios are possible, as are non-rectangular sampling patterns. A single mosaic pointing currently requires ≈ 7.6 s to move the telescope and sample the pointing. Observers must therefore consider the trade-offs between the time required to complete a mosaic, the time scale on which the phenomenon of interest varies in time, and the cadence with which calibrations must be performed.

4. Recent Scientific Results

Despite recent setbacks in telescope access as a result of the pandemic and fierce competition for telescope time since ALMA's return to science operations in late-2021, the publication of ALMA solar science results continues apace, with more than two dozen publications in the past year. Here we highlight just a few recent results. Admittedly, they barely scratch the surface of what ALMA is capable of contributing to chromospheric studies but they convey some of the richness and diversity of ALMA solar science.

4.1. *Calibration of the Brightness Temperature Scale*

Although generally used as a means of filling in the shortest spatial frequencies by combining TP maps with INT data (§3.3), TP maps can be used for science in their own right. A recent example is the work of Alissandrakis et al. (2022), who performed analysis of the center-to-limb brightening of single dish maps in Bands 3, 6, and 7. These authors show that the Eddington-Barbier-like approximation, $T_e(\tau) \approx a_0 + a_1 \ln\tau + a_2(\ln\tau)^2 + a_3(\ln\tau)^3$, can be inverted, yielding $T_B(\mu) \approx A_0 + A_1 \ln\mu + A_2(\ln\mu)^2 + A_3(\ln\mu)^3$, where T_e and T_B are the effective temperature and the brightness temperature, respectively, τ is the optical depth and $\mu = \cos\phi$, with ϕ the heliocentric angle. Furthermore, by using the fact that the H and H^- absorption coefficients are proportional to ν^{-2}, the three frequency bands can be related through $T_B(\mu(\nu/\nu_\circ)^2, \nu_\circ) = T_B(\mu, \nu)$, where ν is a given frequency in Band 3, 6, or 7 and $\nu_\circ$ is a reference frequency. Taking $\nu_\circ = 100$ GHz (Band 3), for which the quiet Sun brightness temperature recommended by White et al. (2017) is $T_B = 7300$ K, all other frequencies and locations on the solar disk can be calibrated against this value.

4.2. *Solar Filaments and Prominences*

A number of recent papers have reported ALMA observations of filaments and prominences, cool material supported in the low corona. Labrosse et al. (2022) observed a prominence in the 3 mm band with ALMA, Hα with the Multi-Chanel Subtractive Double Pass Spectrograph of the Białkow Observatory in Poland, and with SDO/AIA. The morphology of the prominence was very similar in both Hα and at 3 mm. The Hα observations were used to estimate the maximum optical depth of the source at 3 mm ($\tau_{3mm} \approx 2$). This was found to be in good agreement with the observed 3 mm brightness temperature (6000-7000 K) if the prominence material has an effective temperature of ≈ 8000 K, a result in agreement with modeling by Heinzel et al. (2015) and Rodger & Labrosse (2017) and earlier observations (Harrison et al. 1993; Bastian et al. 1993). Comparisons with SDO/AIA EUV data show a good correlation between bright prominence spine material and absorption features, especially in the hot 191Å and 211Å bands. Interestingly, the prominence in the 304Å band is of much greater extent that the 3 mm prominence, indicating the presence of material that is optically thin at 3 mm. da Silva Santos et al. (2022) report sub-arcsecond 1.25 mm (Band 6) observations of an active region filament using mosaicking techniques. Cool filament threads are seen in IRIS Mg II and in the SDO/AIA 304Å passband that have dark counterparts at 1.25 mm, but their visibility varies considerably across the filament spine and as a function of time. Interestingly, contrary to expectations and in contrast to previous observations at lower resolution, observations of the filament at 3 mm show no obvious filament structure (cf. Figure 3). The authors conclude that the filament is not consistent with isothermal conditions. However, the time variability of the filament may also play a role in differences between bands.

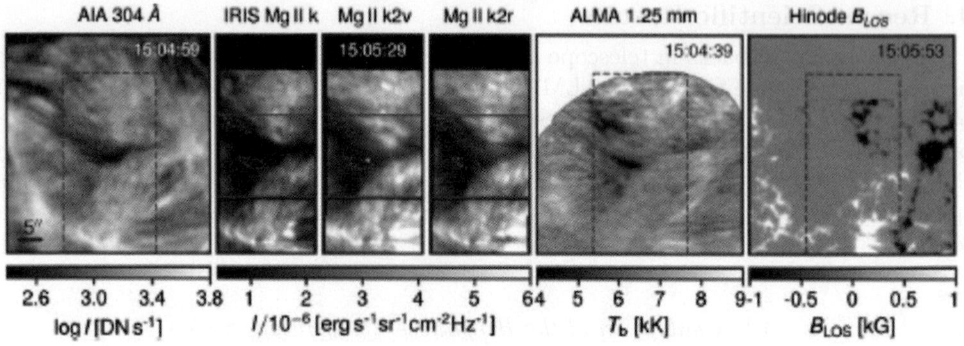

Figure 3. Images of an active region filament. The dashed red line indicates the IRIS field of view. After da Silva Santos et al. (2022).

4.3. *Wave Oscillations in the Chromosphere*

Wave energy in the chromosphere is of paramount importance for understanding energy transfer between, and dissipation within, various layers of the solar atmosphere. Several papers have appeared in the past year that present preliminary assessments of wave phenomena in the chromosphere from various perspectives. Chai et al. (2022) focus on oscillations over a sunspot, reporting the first high cadence ALMA 3 mm observations of 3 min brightness temperature oscillations in concert with Hα observations by the Goode Solar Telescope as well as comparisons with IRIS UV and SDO/AIA EUV data. The 3 mm oscillations observed by ALMA are confined to the sunspot umbra. An examination of the relative phasing of the ALMA oscillations with those in the red and blue wings of Hα are consistent with those expected for heating by a propagating disturbance. Narang et al. (2022) took a broader view, exploring the distribution of oscillation power in a plage region using ALMA observations at 1.25 mm (Band 6) and comparing them with IRIS and SDO/AIA observations. They find no significant correlation between 1.25 mm oscillations (distributed over a broad range of periods) and those seen in the UV/EUV bands (predominenantly periods of 4-7 min). They suggest the lack of association between ALMA 1.25 mm oscillations and those in UV/EUV bands may be the result of different formation heights. Finally, Guevara Gómez et al. (2022) report an ALMA 3 mm study of high frequency oscillations in compact bright chromospheric features, with periods ranging from $\sim 1-2$ min. On the basis of an anti-correlation between 3 mm brightness temperature variations and the size of the compact chromospheric features, the authors suggest that it may be indicative of sausage-mode waves in these structures.

4.4. *Chromospheric Microflares*

While observations of solar flares remains a future capability, it is already possible to observe small transient events since they do significantly increase the system temperature. The role of microflares and nanoflares in the chromospheric and coronal heating budgets of active regions is of great interest. Shimizu et al. (2021) report joint observations of a "chromospheric microflare" with ALMA, IRIS, and *Hinode*/XRT. The thermal energy content of the X-ray microflare was estimated to be a few $\times 10^{26}$ ergs. Counterparts to the soft X-ray microflare were observed in the loop footpoints of the microflare SXR-emitting loop by ALMA at 3 mm and by IRIS in the Si IV line that were similar to the signature of the Neupert effect. Interestingly, no clear counterparts were observed in the Mg II h and k lines, and only a marginal signature was seen in the C II line. The thermal energy content of the 100 GHz footpoint sources was two orders of magnitude less than that estimated for X-ray emitting loop. The authors suggest that the footpoint signatures are the result of nonthermal electrons impinging on the upper chromosphere. Abe et al. (2022) have studied even tinier events using ALMA at 3 mm, those with a thermal energy of only 10^{20-22} ergs. They associated these tiny events with a small emerging magnetic flux region and with a magnetic neutral line. These results can be compared with those of Nindos et al. (2020) who found weak transients throughout the quiet chromosphere with thermal energies of $\sim 10^{24-26}$ ergs. There is considerable uncertainty in these energy estimates and much work remains to characterize the energy distribution of microflares and nanoflares in the solar chromosphere and their role in chromospheric heating.

To conclude this section we note that ALMA science data is available through the ALMA science archive. After a proprietary period of one year, all ALMA data are publicly available for further analysis. A far more convenient means of accessing ALMA archival data, however, is through the Solar ALMA Science Archive (SALSA) hosted by the Science Data Center at the University of Oslo, as described by Henriques et al. (2022). SALSA currently contains 26 "science ready" datasets, an imaging pipeline, a library of auxiliary tools, and links to other datasets from other observatories.

5. Future Capabilities

There are many more ALMA capabilities that can be implemented. The rate at which this can be done is limited to available manpower and resources. Among those likely to be tested and commissioned in the next 1-3 years are the following.

5.1. *Polarimetry*

Efforts are currently underway to implement polarimetric observations with ALMA. The Stokes V parameter is of primary interest since it can be used to constrain the line-of-sight magnetic field in active regions. Modeling by Loukitcheva et al. (2017) suggests that the fractional circular polarization will be highest in the 3 mm band. Hence initial efforts have been directed toward Band 3. Polarimetry is expected to be made available to the community in 2023 or 2024.

5.2. *Flare Mode*

Observations of flares at mm/sub-mm-λ will offer fundamentally new insights into energy release, particle acceleration, and emission mechanisms on the Sun. particularly the mysterious "sub-THz" spectral component review by Krucker et al. (2013). Implementation of flare mode observing with ALMA poses a number of challenges, not the least of which managing the high signal levels expected. The current scheme briefly described in §1 used to observe the quiet Sun is not appropriate for flare observations. It is likely that special solar filters on each antenna will be used to attenuate the incident signal. However, the robust calibration of solar flare observations with ALMA, as well as a host of operational issues, must be also be addressed before offering the mode to the community. An ALMA Solar Development Study proposal is under consideration to provide the necessary resources to enable this mode.

5.3. *Spectral Line Mode*

To date, ALMA has provided continuum observations in four frequency bands. In practice, each band is divided into four spectral windows of 2 GHz bandwidth and each spectral window is channelized into 128 spectral channels. After calibration and data flagging, current practice is to average over each spectral window to form pseudo-continuum datasets. It has been know for may years that H radio recombination lines

(RRLs) are present in the mm/sub-mm band. Previous searches have generally failed to detect them – they are expected to be very broad and shallow at a result of pressure broadening. The only reported detections are those of Clark et al. (2000a,b), who reported detections of the H21α (662.4 GHz) and H19α (888 GHz) RRLs with a Fourier Transform Spectrometer. The H26α RRL in Band 6 is an obvious target for the much more sensitive ALMA telescope. Other potential lines include CO rotational transitions (Wedemeyer-Böhm & Steffen 2007). While CO has been detected in infrared lines (Ayres et al. 2006) no detections have yet been made in the mm/sub-mm-λ band. The detection of such broad and shallow lines will require careful bandpass calibration and continuum subtraction. Science verification of their detectability is needed before spectral line mode observing is implemented, however.

5.4. *Additional Frequency Bands*

Additional frequency bands can be offered in the coming years. An exciting prospect is the availability of ALMA Band 1 (35-50 GHz) and Band 4 (125-163 GHz), thereby providing frequency coverage over a decade of bandwidth (or wavelengths from 0.8-8 mm), allowing the solar atmosphere to be probed from the low to upper chromosphere. Band 1 is a new capability (Huang et al. 2016), not just for the solar community, but for the community at large. It is expected to be available to the astronomical community in Cycle 10, but support for solar observing will likely be another year or two later. While solar observing at higher frequency bands is under consideration (Bands 8, 9, and 10) they offer special challenges, both technical and operational. High frequency interferometric observations require exceptional weather conditions, conditions rarely achieved during daytime. It may be possible to observe with these bands using the TP antennas (Band 9 TP mapping has already been demonstrated), but that will require policy changes.

5.5. *Higher Angular Resolution*

While there are limits imposed by the imaging capabilities of a fixed configuration of antennas as discussed in §3.2, these limits are not well understood. Higher angular resolution is a goal that is only achievable by observing with larger ALMA array configurations and/or higher frequency bands. An ESO-funded ALMA Development Study is assessing strategies for achieving higher angular resolution imaging of the Sun with ALMA is underway.

6. Concluding Remarks

ALMA is a remarkable instrument that has opened a new window onto solar physics. Access to the telescope, in competition with the wider astronomical community, and its use for solar observing is not without challenges. Nevertheless, ALMA brings fundamentally new capabilities to solar observations: continuum imaging of the Sun at mm/sub-mm-λ with high time resolution and with arcsecond angular resolution. Continuing development will expand its capabilities even further. With the availability of new instruments such as the Daniel K. Inouye Solar Telescope operating at O/IR wavelengths, powerful new synergies will be possible that will increase the impact both instruments on outstanding problems in solar physics.

Acknowledgements

I thank my many colleagues on the ALMA Solar Development team and members of the JAO staff for their efforts in making solar observing with ALMA a reality: M. Bárta (Czech ARC), R. Brajša (Zagreb), B. Chen (NJIT), B. De Pontieu (LMSAL),

G. Fleishman (NJIT), D. Gary (NJIT), A. Hales (JAO), A. Hirota (JAO), H. Hudson (Glasgow), R. Hills (MRAO), S. Ishii (NAOJ), K. Iwai (Nagoya), G. Motorina (Czech ARC), N. Phillips (ESO), M. Shimojo (NAOJ), I. Skokić (Czech ARC), S. Wedemeyer (Oslo), S. White (USAF), P. Yagoubov (ESO), Y. Yan (NAO/CAS). The National Radio Astronomy Observatory is a facility of the National Science Foundation operated under cooperative agreement by Associated Universities, Inc.

References

Abe, M., Shimizu, T., & Shimojo, M. 2022, Frontiers in Astronomy and Space Sciences, 9, 908249
Alissandrakis, C. E., Bastian, T. S., & Nindos, A. 2022, A&A, 661, L4
Ayres, T. R., Plymate, C., & Keller, C. U. 2006, ApJS, 165, 618
Bastian, T., Shimojo, M., Barta, M., White, S., & Iwai, K. 2022, arXiv e-prints, arXiv:2209.01659
Bastian, T. S., Ewell, M. W., J., & Zirin, H. 1993, ApJ, 415, 364
Chai, Y., Gary, D. E., Reardon, K. P., & Yurchyshyn, V. 2022, ApJ, 924, 100
Clark, T. A., Naylor, D. A., & Davis, G. R. 2000a, A&A, 357, 757
Clark, T. A., Naylor, D. A., & Davis, G. R. 2000b, A&A, 361, L60
da Silva Santos, J. M., White, S. M., Reardon, K., et al. 2022, Frontiers in Astronomy and Space Sciences, 9, 898115
Guevara Gómez, J. C., Jafarzadeh, S., Wedemeyer, S., & Szydlarski, M. 2022, A&A, 665, L2
Harrison, R. A., Carter, M. K., Clark, T. A., et al. 1993, A&A, 274, L9
Heinzel, P., Berlicki, A., Bárta, M., Karlický, M., & Rudawy, P. 2015, Sol. Phys, 290, 1981
Henriques, V. M. J., Jafarzadeh, S., Guevara Gómez, J. C., et al. 2022, A&A, 659, A31
Huang, Y. D. T., Morata, O., Koch, P. M., et al. 2016, in Society of Photo-Optical Instrumentation Engineers (SPIE) Conference Series, Vol. 9911, Modeling, Systems Engineering, and Project Management for Astronomy VI, ed. G. Z. Angeli & P. Dierickx, 99111V
Krucker, S., Giménez de Castro, C. G., Hudson, H. S., et al. 2013, AApR, 21, 58
Labrosse, N., Rodger, A. S., Radziszewski, K., et al. 2022, MNRAS, 513, L30
Linsky, J. L. 1973, Sol. Phys, 28, 409
Loukitcheva, M., White, S. M., Solanki, S. K., Fleishman, G. D., & Carlsson, M. 2017, A&A, 601, A43
Narang, N., Chandrashekhar, K., Jafarzadeh, S., et al. 2022, A&A, 661, A95
Nindos, A., Alissandrakis, C. E., Patsourakos, S., & Bastian, T. S. 2020, A&A, 638, A62
Rodger, A. & Labrosse, N. 2017, Sol. Phys, 292, 130
Shimizu, T., Shimojo, M., & Abe, M. 2021, ApJ, 922, 113
Shimojo, M., Bastian, T. S., Hales, A. S., et al. 2017, Sol. Phys, 292, 87
Taylor, G. B., Schwab, F. R., & Perley, R. A., eds. 1999, Synthesis Imaging in Radio Astronomy II (PASP)
Vernazza, J. E., Avrett, E. H., & Loeser, R. 1981, ApJS, 45, 635
Wedemeyer-Böhm, S. & Steffen, M. 2007, A&A, 462, L31
White, S. M., Iwai, K., Phillips, N. M., et al. 2017, Sol. Phys, 292, 88
Wootten, A. & Thompson, A. R. 2009, IEEE Proceedings, 97, 1463

The Era of Multi-Messenger Solar Physics
Proceedings IAU Symposium No. 372, 2023
G. Cauzzi & A. Tritschler, eds.
doi:10.1017/S1743921323000418

Modeling efforts for multi-mission science

C. Sasso[1] and A. P. Rouillard[2]

[1]National Institute for Astrophysics, Astronomical Observatory of Capodimonte, Salita Moiariello 16, I-80131 Napoli, Italy

[2]Institut de Recherche en Astrophysique et Planétologie, CNRS, UPS, CNES, Toulouse, France

Abstract. The Solar Orbiter spacecraft, launched in February 2020, is equipped with both remote-sensing (RS) and in-situ (IS) instruments to record novel and unprecedented measurements of the solar atmosphere and the inner heliosphere. To take full advantage of these new datasets, we have developed tools and techniques to facilitate multi-instrument and multi-spacecraft studies. In particular the yet inaccessible low solar corona below 2 $R_\odot$ can only be observed remotely and techniques must be used to retrieve coronal plasma properties in time and in 3-D space. These properties are useful to drive numerical models and test the different theories proposed to describe the fundamental processes of the solar atmosphere. In addition, the last decades of research have shown that the coupling between the solar corona and the heliosphere is most efficiently studied by combining RS with IS data. During one of the last Solar Orbiter remote sensing windows (March 2022), planned for the Solar Orbiter instruments, we ran complex observation campaigns to maximize the likelihood of linking IS data to their source region near the Sun, by directing some RS instruments to specific targets on the solar disk just days before data acquisition. We show how it is possible to achieve these results directed to improve our understanding of how heliospheric probes connect magnetically to the solar disk.

Keywords. Sun: atmosphere, Sun: magnetic fields, Sun: corona, Sun: activity

1. Introduction

The Solar Orbiter (SolO) mission (Müller et al. 2020) is equipped with a unique dedicated payload of 10 remote-sensing (RS) and in-situ (IS) instruments measuring from the photosphere into the solar wind, designed to observe the Sun and the heliosphere.

One of the SolO's main science goal is to establish the link between the RS observations and the IS measurements in order to link heliospheric phenomena back to their sources on the Sun.

This task is very challenging to achieve since it requires the availability of good coronal and heliospheric models of the sun's magnetic field configuration that can be run very quickly, ingesting daily information from the IS instruments to inform promptly the spacecraft about the pointing of the RS instruments. Robust, stable operational models able to establish the most likely connection between the spacecraft and the solar surface are therefore needed but they do not need to be necessarily the most complex and/or sophisticated scientific ones.

In order to take care of this issue, i.e., coordinating modeling and theoretical support for SolO science to be able to choose targets and update the pointing, the European Space Agency (ESA) established the Modeling and Data Analysis Working Group (MADAWG) for the Solar Orbiter mission.

The MADAWG, during the years, had also to take care of the following topics:

- Defining the format of SolO instrument datasets and their metadata;
- Defining how SolO instruments data archives will exchange SolO data;
- Preparing data analysis tools to relate different SolO datasets.

An exhaustive description of the community-led effort guided by the MADAWG to develop models, tools, and techniques in order to test different theories to explain the physical processes that occur in the solar plasma, can be found in the Rouillard et al. (2020) paper. The present contribution can be considered a brief update of it, since it reviews, in particular, the effort made during the last two remote sensing windows of the SolO instruments (March and October 2022), to direct the pointing of the RS instruments, starting from the modeling results combined with the RS and IS data, to maximise the likelihood of linking IS data to their source region near the Sun.

Several RS instruments on board SolO, indeed, can be directed to specific targets situated on the solar disk just days before data acquisition, enabling coordinated campaigns to point to the predicted source region of different phenomena (like slow/fast solar wind), that will be measured by IS payload at time of arrival at spacecraft.

2. Magnetic Connectivity Tool

In order to forecast where to point the RS instruments, we need to apply models based on coronal magnetic field extrapolations and different techniques to retrieve coronal plasma properties in time and in three dimensional (3D) space. This is because we are not able to measure directly the Sun's magnetic field in the corona.

To predict the connectivity points on the solar disk (i. e., the footpoints of magnetic field lines connected to a spacecraft, SolO in our case), the MADAWG selected the use of one particular tool, the Magnetic Connectivity tool (`http://connect-tool.irap.omp.eu/`, operational since 2018, Rouillard et al. 2020).

Figure 1 shows the main page of the Magnetic Connectivity Tool web interface that enables the operator to choose different combinations of parameters to estimate the connectivity points close to Sun's surface. There is the possibility to chose how to model the inner solar corona ("coronal magnetic field" keyword) and the tool also enables the inclusion of forward or backward propagation at solar wind speed or speed of light ("propagation mode" keyword). The list at the bottom of the page gives the possibility to select the preferred vantage point (Earth or a S/C).

The SolO RS windows campaigns performed in March and October 2022, took advantage of the Magnetic Connectivity Tool to forecast the pointing of the spacecraft and all the extrapolations assumed the same model that we are describing in the following. The idea at the base of the assumed magnetic field model is to use the simplest approach, i.e. to describe the interplanetary magnetic field by a Parker spiral up to a certain distance in the corona and below that height (inner corona) to describe the field by using a potential field source surface (PFSS). The different approach at different heights in the solar atmosphere, is needed because in the inner corona we have open field lines that are deviated by closed magnetic field lines (see Fig. 2). The PFSS model (Altschuler & Newkirk 1969; Schatten et al. 1969), in particular, computes the magnetic field between 1 and 2.5 $R_{\odot}$, by extrapolation from the measured photospheric magnetic field, assuming that the field is current free and radial at 2.5 $R_{\odot}$. The photospheric magnetic field that consists in the boundary condition for the coronal extrapolations is derived from Air Force Data Assimilative Photospheric Flux Transport (ADAPT) synoptic maps of the radial component of the magnetic field (Arge et al. 2010; Hickmann et al. 2015), that are in turn constructed from Global Oscillation Network Group (GONG;

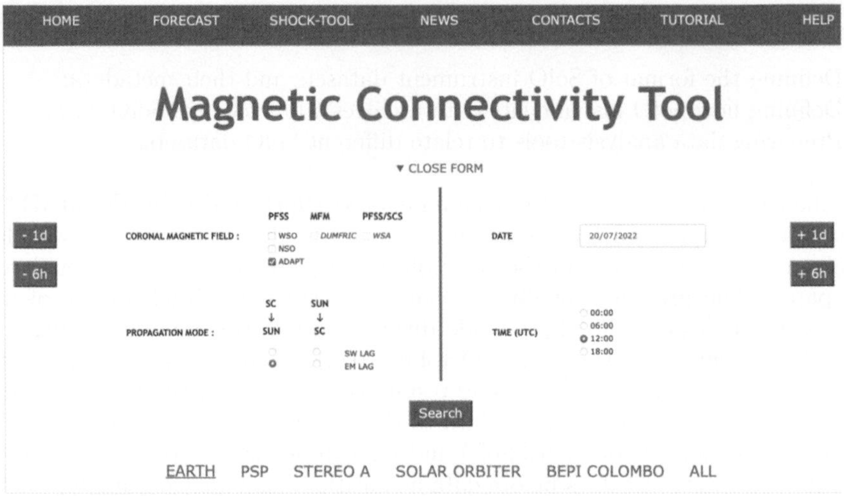

Figure 1. Interface of the Magnetic Connectivity Tool (`http://connect-tool.irap.omp.eu/`).

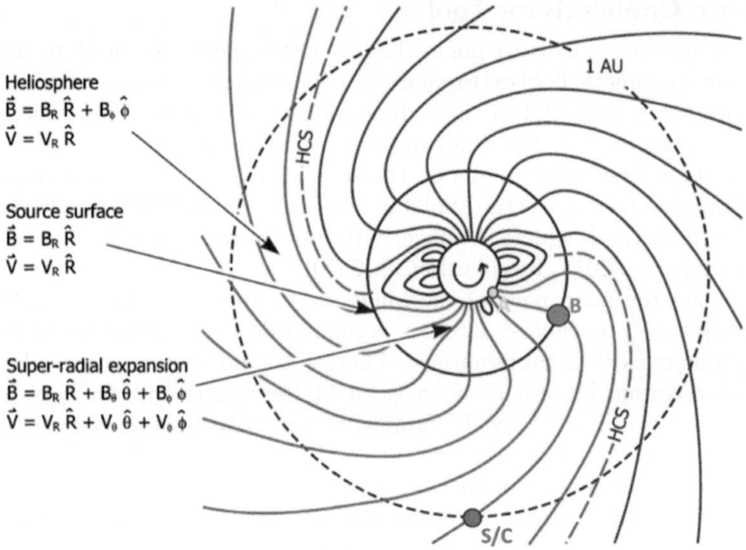

Figure 2. Schematic (from Owens & Forsyth 2013) illustrating how magnetic connectivity is established in the Magnetic Connectivity Tool, showing the location of a spacecraft (S/C), the intersection of the Parker Spiral (B) and the photospheric/low corona magnetic footpoint of field line (A).

Hill et al. 1994, 1996) magnetograms by evolving them using a photospheric flux transport model (Worden & Harvey 2000). Therefore, ADAPT generates global solar photospheric magnetic field maps using flux transport providing synchronic solutions every 24 hours. The result of this extrapolation are 12 maps for each predicted time. The connectivity tool is equipped with an automated evaluation of PFSS reconstructions by combining the position of the Heliospheric Current Sheet (HCS or neutral line) with the position of the streamers in the coronagraphic white light maps (Poirier et al. 2021). Currently, the code is exploiting the SOlar and Heliophysics Observatory/LASCO-C2 -C3 (SOHO, Domingo et al. 1995) maps in real time and a combination of the maps

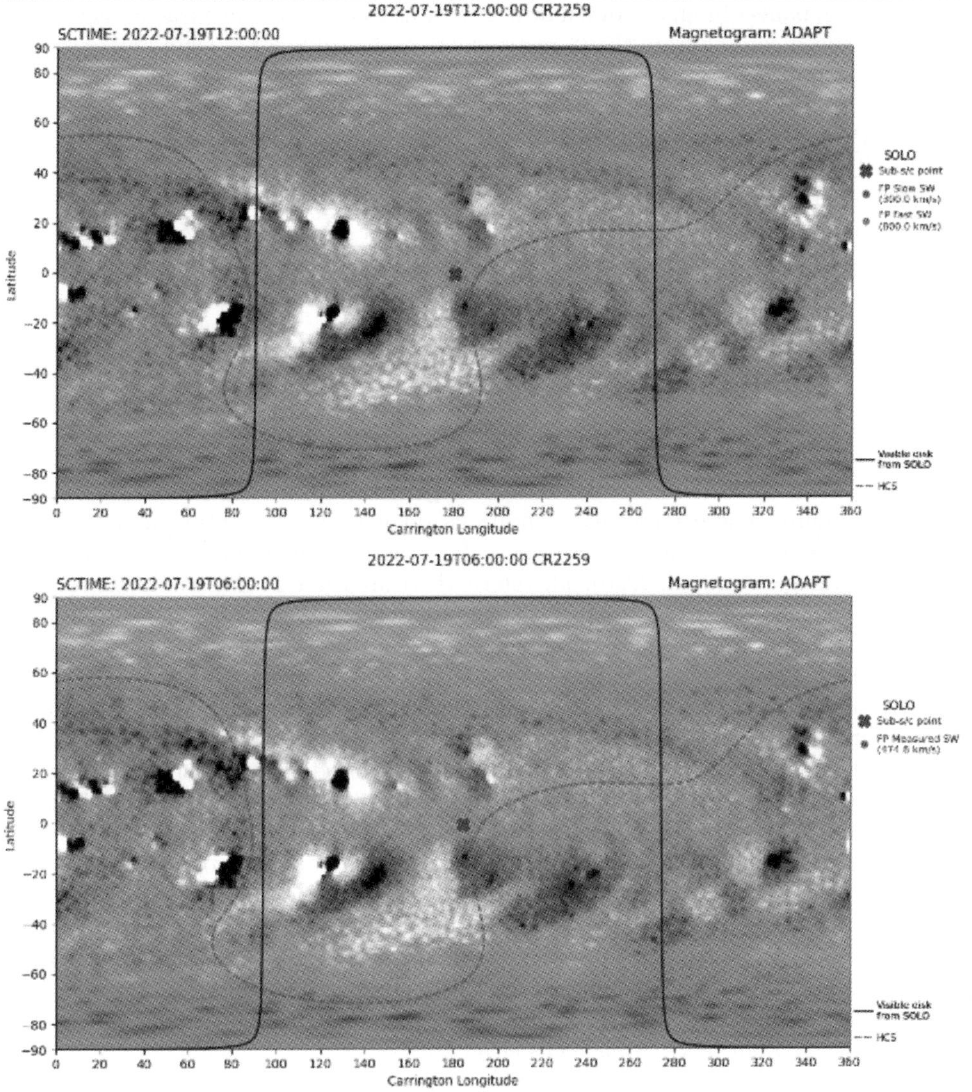

Figure 3. Examples of a connectivity map estimated without (top panel) and with (bottom) the integration of the measured solar wind speed from the SWA/PAS LLD. The red line is the HCS while the green and red dots represent the fast and slow wind connection points, respectively.

obtained by combining the data from SOHO/LASCO-C2 and Solar-TErrestrial RElations Observatory/SECCHI–COR1 (STEREO, Kaiser et al. 2008) for post-analysis, following the example of the work of Sasso et al. (2019). There is on-going work to integrate in the magnetic connectivity tool pipeline also SolO/Metis (Antonucci et al. 2020) Low Latency Data (LLD) and to use a combination of magnetograms, including SolO/Photospheric and Helioseismic Imager (PHI, Solanki et al. 2020) ones.

The top panel of Fig. 3 gives an idea of a connectivity map forecasted by the connectivity tool using ADAPT/GONG magnetic field maps with an estimate for the speed of the wind stream reaching the S/C (SolO on this example). The red line gives and estimation of the position of the HCS while the green and red dots represent the forecasted positions of the fast and slow wind connection points, respectively.

The bottom panel of Fig. 3 shows how it's possible to integrate the measurements of the solar wind speed coming from the IS instruments on board SolO to get a better estimate of the position of connection points. At the time it was created the map in the bottom panel of Fig. 3, the tool found information on the measured solar wind speed coming from the Proton-Alfa Sensor (PAS) LLD of the SWA instrument (Owen et al. 2020) on-board SolO. For this reason, in the bottom map, the tool gives only one estimation of the solar wind speed (red dots) and not two (green and red dots) as in the top panel of the same figure. Indeed, the connectivity tool has been implemented with a pipeline that automatically downloads SWA/PAS LLD data when they are available. SolO LLD data are fundamental to help establishing the most likely source of plasma in these maps.

3. SolO RS instruments campaign

SolO is a more likely a planetary mission with its typical operational constraints such as limited available telemetry due to the characteristics of the orbits, variable data latency due to the variations of the distance of the spacecraft from Earth, and limited data storage on-board. These particular conditions do not permit continuous observations for the RS instruments and force the need for long-term planning of top-level science operations (Zouganelis et al. 2020). Therefore, there are three 10-day windows every orbit (six months long), called the remote sensing windows (RSWs), typically centred around perihelia and minimum and maximum heliographic latitudes, where RS instruments can run their high resolution campaigns. Outside the RSWs, the RS instruments perform synoptic observations monitoring the Sun at low data acquisition rate, while the IS instruments are operating continuously along the orbit.

During the RSWs, there is the possibility to plan coordinated observations involving the SolO instruments with the possibility of deciding the spacecraft pointing just some days in advance. The science planning foresees what is called a very short term planning (VSTP) of science operations, inside the long-term plan (LTP, programmed 3 months before the observations, was formerly 6 months) and the short-term plan (STP, 1 week before). This possibility is there since high-resolution science requires fine-pointing to a target which position cannot be pre-planned, being solar activity unpredictable so much time in advance. During the planning of the LTP, the observing campaigns, called SOOPs for Solar Orbiter Observing Plans, are inserted and consists in the building blocks of the Science Activity Plan (SAP, Zouganelis et al. 2020).

During the latter VSTPs campaigns (March and October 2022), the MADAWG, the SolO Science Operations Centre (SOC), and the SOOP coordinators were working in close contact to forecast the location of targets and their magnetic connection with the S/C to choose RS targets and update the pointing. A big effort was done by the IS community to provide LLD that, as we described in Sect. 2, have an important role in the connectivity tool, helping retrieving the most reliable magnetic connection with the spacecraft.

At the time we are writing this contribution, first results are coming from two "Slow Wind Connection" SOOP campaigns held in the period March 3–6, and March 17–22, 2022. Both SOOPs were aiming at catching, with the RS instruments, the dynamics of an open-closed field boundary, while the wind associated with such a boundary would then be measured in situ. The target of the SOOPs were, therefore, active region/coronal hole boundaries and the science goal was to understand what are the release mechanisms of the slow solar wind at an open-closed magnetic field boundary.

The campaign consisted in coordinated observations involving the following SolO instruments: spectra from the Spectral Imaging of the Coronal Environment (SPICE, SPICE Consortium et al. 2020), 1 hr/day high resolution Extreme Ultraviolet

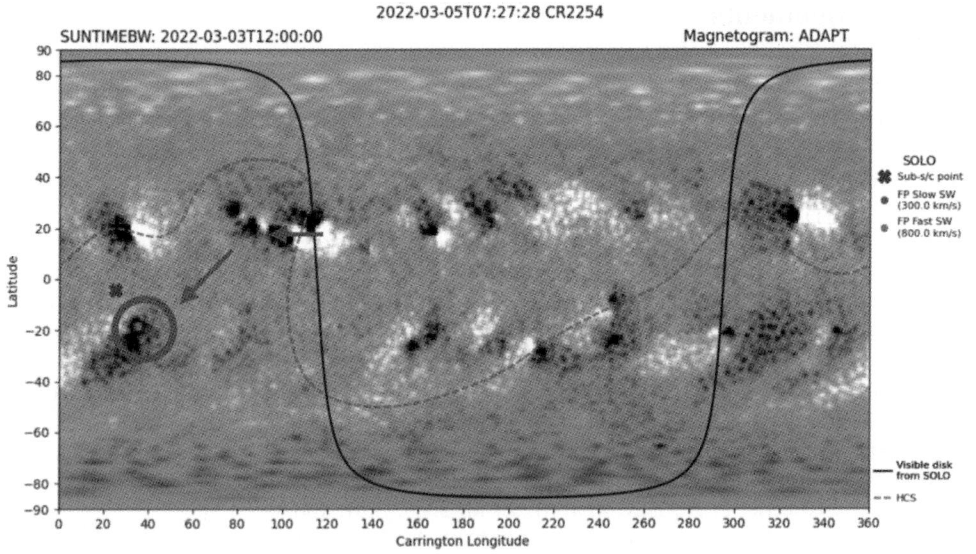

Figure 4. Connectivity map forecasted three and two days before the planned observation. The red point outside the red circle indicates the slow wind connection point forecasted for the 2nd of March, while the red point inside the red circle indicates the slow wind connection point forecasted for the 3nd of March. As in Fig. 3, the red line is the HCS while the green dots represent the fast wind connection points.

Imager (EUI, Rochus et al. 2020) images and PHI magnetograms, and continuous EUI/PHI/STIX (Spectrometer/Telescope for Imaging X-rays, Krucker et al. 2020)/SolOHI (Heliospheric Imager, Howard et al. 2020)/Metis synoptics and in-situ measurements.

Figure 4 shows an example of a connectivity map forecasted during these SOOPs. In particular, it shows the connectivity map obtained two days (3rd of March) before the planned observation, the 5th of March. On the map we have also superimposed the slow wind connection point forecasted for the 2nd of March (the red point outside the red circle). We see that connectivity moves from northern to southern active region (AR) during one day and this episode underlines the importance of having the forecasting and the decision on the pointing done as close as possible to the observations' day. After consulting the maps, SolO was predicted to be connected to the southern AR that was just rotating into spacecraft view on the 3rd of March.

Preliminary results coming from both observing campaigns, currently analyzed by a group coordinated by the SOOPs coordinator, S. Yardley, show that slow wind was measured at the SolO spacecraft during both windows.

4. Conclusions

We described the on-going effort by ESA's MADAWG to provide data-analysis tools and models that can support mission operations as well as help improving SolO science outputs with a particular focus on tools that exploit the coordinations between RS and IS instruments and the comparison of their data.

The preliminary results coming from the first coordination campaigns involving the majority of instruments on-board SolO aimed at forecasting magnetic connectivity between the particles arriving at the IS instruments and their sources back to the Sun, by using MADAWG's tools, are very encouraging and we are sure our work will help improve our understanding of the solar magnetic corona and heliosphere.

Acknowledgements

Solar Orbiter is a joint ESA and NASA mission. The authors would like to thank the editors, G. Cauzzi and A. Tritschler, for the invitation and C.S. would like to thank the IAU for travel support to attend the IAUGA2022 in South Korea, in person.

References

Altschuler, M. D. & Newkirk, G. 1969, Sol. Phys, 9, 131

Antonucci, E., Romoli, M., Andretta, V., et al. 2020, A&A, 642, A10

Arge, C. N., Henney, C. J., Koller, J., et al. 2010, in American Institute of Physics Conference Series, Vol. 1216, Twelfth International Solar Wind Conference, ed. M. Maksimovic, K. Issautier, N. Meyer-Vernet, M. Moncuquet, & F. Pantellini, 343–346

Domingo, V., Fleck, B., & Poland, A. I. 1995, Sol. Phys, 162, 1

Hickmann, K. S., Godinez, H. C., Henney, C. J., & Arge, C. N. 2015, Sol. Phys, 290, 1105

Hill, F., Fischer, G., Grier, J., et al. 1994, Sol. Phys, 152, 321

Hill, F., Stark, P. B., Stebbins, R. T., et al. 1996, Science, 272, 1292

Howard, R. A., Vourlidas, A., Colaninno, R. C., et al. 2020, A&A, 642, A13

Kaiser, M. L., Kucera, T. A., Davila, J. M., et al. 2008, SSR, 136, 5

Krucker, S., Hurford, G. J., Grimm, O., et al. 2020, A&A, 642, A15

Müller, D., St. Cyr, O. C., Zouganelis, I., et al. 2020, A&A, 642, A1

Owen, C. J., Bruno, R., Livi, S., et al. 2020, A&A, 642, A16

Owens, M. J. & Forsyth, R. J. 2013, Living Reviews in Solar Physics, 10, 5

Poirier, N., Rouillard, A. P., Kouloumvakos, A., et al. 2021, Frontiers in Astronomy and Space Sciences, 8, 84

Rochus, P., Auchère, F., Berghmans, D., et al. 2020, A&A, 642, A8

Rouillard, A. P., Pinto, R. F., Vourlidas, A., et al. 2020, A&A, 642, A2

Sasso, C., Pinto, R. F., Andretta, V., et al. 2019, A&A, 627, A9

Schatten, K. H., Wilcox, J. M., & Ness, N. F. 1969, Sol. Phys, 6, 442

Solanki, S. K., del Toro Iniesta, J. C., Woch, J., et al. 2020, A&A, 642, A11

SPICE Consortium, Anderson, M., Appourchaux, T., et al. 2020, A&A, 642, A14

Worden, J. & Harvey, J. 2000, Sol. Phys, 195, 247

Zouganelis, I., De Groof, A., Walsh, A. P., et al. 2020, A&A, 642, A3

Clementina Sasso from INAF (Italy) talking about "magnetic connectivity tools" to investigate the relationship between in-situ measurements and their source regions on the Sun.

Chapter II:
MAGNETIC FIELDS - GENERATION, ACTIVITY, AND INSTABILITIES

The IAUS372 audience listening to the review talk by T. Li (China) on CMEs and flares, August 2, 2022.

The Era of Multi-Messenger Solar Physics
Proceedings IAU Symposium No. 372, 2023
G. Cauzzi & A. Tritschler, eds.
doi:10.1017/S1743921322004665

Solar flare-CME association

Ting Li[1,2]

[1]National Astronomical Observatories, Chinese Academy of Sciences, Beijing 100101, China; liting@nao.cas.cn

[2]School of Astronomy and Space Science, University of Chinese Academy of Sciences, Beijing 100049, China

Abstract. We carry out the first statistical study that investigates the flare-coronal mass ejections (CMEs) association rate as function of the flare intensity and the total unsigned magnetic flux (Φ_{AR}) of ARs that produces the flare. Our results show that flares of the same GOES class but originating from an AR of larger Φ_{AR}, are much more likely confined. This implies that Φ_{AR} is a decisive quantity describing the eruptive character of a flare, as it provides a global parameter relating to the strength of the background field confinement. We also calculated the mean twist values α in regions close to the polarity inversion line and proposed a new parameter $\alpha/\Phi_{\rm AR}$ to measure the probability for a large flare to be associated with a CME. We find that the new parameter $\alpha/\Phi_{\rm AR}$ is well able to distinguish eruptive flares from confined flares.

Keywords. Sun: activity, Sun: flares, Sun: coronal mass ejections (CMEs), Sun: magnetic fields

1. Introduction

Solar flares are sudden, localized brightenings of the solar atmosphere evident throughout the entire electromagnetic spectrum and often occur in association with coronal mass ejections (CMEs). During these solar activities, a vast amount of magnetic energy is released, converted into particle acceleration, kinetic energy and plasma heating (Fletcher et al. 2011). According to whether solar flares are associated with CMEs, they can be divided into two types, "eruptive flares" associated with a CME and "confined flares" not associated with a CME. Statistically, the percentage of flares associated with CMEs increases with the strength of the flare, with the strongest X-class flares almost always associated with a CME (Yashiro et al. 2006). Understanding the association between flares and CMEs is important to forecast space weather in the near-Earth environment and meanwhile, the solar flare-CME paradigm might be applied to magnetic activities in other stars, which is vital for the question of exoplanet habitability and the evolution of stellar mass loss and rotation (Lammer et al. 2007).

Two factors are typically suggested to define whether a flare would be eruptive or confined. The first factor describes the active region (AR) non-potentiality, i.e., magnetic helicity, free magnetic energy, twist, etc. (Liu et al. 2016; Toriumi et al. 2017). Recently, Gupta et al. (2021) found that the AR is likely to produce large CME-associated flares if the relative contribution of free magnetic energy, the fraction of non-potential helicity and the normalized current-carrying helicity are larger than certain thresholds. Avallone & Sun (2020) and Kazachenko et al. (2022) found that confined ARs tend to be more current neutralized than eruptive ARs. The second factor describes the constraining effect of the overlying field: its decay rate with height and strength (Sun et al. 2015; Amari et al. 2018). Wang & Zhang (2007) and Baumgartner et al. (2018) found that confined events often occur closer to the magnetic center of ARs, while eruptive events tend to occur close to the edge of ARs. The decay index, which is derived from magnetic field extrapolations, describes how fast the constraining magnetic field decays and, thus, determines whether a full eruption could take place or not (Török & Kliem 2005). A faster decrease, i.e., higher decay index, reflects more favourable conditions for eruptive flares (Wang et al. 2017; Vasantharaju et al. 2018).

2. Summary of our results

Based on the *Solar Dynamics Observatory* (*SDO*; Pesnell et al. 2012) observations, we established a database including a total of 322 flares (170 eruptive and 152 confined) of Geostationary Operational Environmental Satellite (*GOES*) class M1.0 and larger that occurred within 45° from the central meridian, from June 2010 until June 2019 (see detailed description in Li et al. 2020). For each event, we obtained these GOES parameters from the GOES soft X-ray (SXR) (1-8 Å channel) light curve: 1. FWHM of the light curve, Duration (FWHM); 2. e-folding decay time, Duration (decay); and 3. peak flux, F_{SXR}. Duration (decay) was obtained from the time derivative of the GOES 1-8 Å flux, and defined as the period when the derivative is above 1/e of the maximum value (Song & Tian 2018).

The AR parameters are obtained by using the available vector magnetograms from Space-Weather HMI AR Patches (Bobra et al. 2014) before the flare onset. Total AR magnetic flux, AR area and normalized field strength, which is the total AR flux normalized by the AR area, are calculated for each event by considering all pixels of $|B_r|$>100 G. Four ribbon parameters are calculated for each event, including the ribbon area, total unsigned flux inside the ribbon (obtained from the RibbonDB catalog in Kazachenko et al. 2017), normalized field strength of the ribbon and ribbon distance. In Figure 1, we give an example of a confined M2.7-class flare on 2013 October 23, which initiated at 20:41 UT and peaked at 20:53 UT. We extracted the flare ribbons from 1600 Å images (Figures 1(a)-(b)) by defining them as the pixels with intensity at any time during the flaring (from the flare start time to the end time) equal to or larger than 40σ (standard deviation) above the mean of the quiet-Sun values, and we made composite AIA 1600 Å flare ribbons (Figure 1(c)). The separation between the two area-weighted centroids of the composite ribbons in the positive and negative polarities is obtained as the parameter of ribbon distance (Figure 1(d)). The FWHM duration of the GOES light curve for this event is about 670 s (Figures 1(e)-(f)).

Figure 2 displays the histograms of GOES parameters, AR parameters and ribbon parameters for confined (red lines) and eruptive (blue lines) events. As seen from Figures 2(a1)-(a2), there are slight differences in distributions of Duration (FWHM) and Duration (decay) between the confined and eruptive cases. The averages of the log values of Duration (FWHM) are 652 s (indicated by vertical dotted red line in panel (a1)) and 772 s (vertical dotted blue line) for confined and eruptive cases (difference=17%, hereafter we use relative difference |a-b|/(|a+b|/2) to show the quantitative difference between a and b), respectively. The averages of the log values of Duration (decay) are 1122 s (indicated by vertical dotted red line in panel (a2)) and 1360 s (vertical dotted blue line) for confined and eruptive cases (difference=19%), respectively. The histogram of GOES peak flux F_{SXR} shows that the number of confined flares is more than eruptive flares for <M2.0-class flares and that of confined flares is less than eruptive flares for >M2.0-class (panel (a3)).

For AR parameters, there are significant differences in distributions of AR magnetic flux and AR area between the confined and eruptive cases (panels (b1) and (c1)). The confined events have larger AR magnetic flux and AR area. The averages of the log values of Φ_{AR} (indicated by vertical dotted lines) are 6.5×10^{22} Mx and 4.4×10^{22} Mx for confined and eruptive cases (difference=39%), respectively. The log-mean values of AR area for

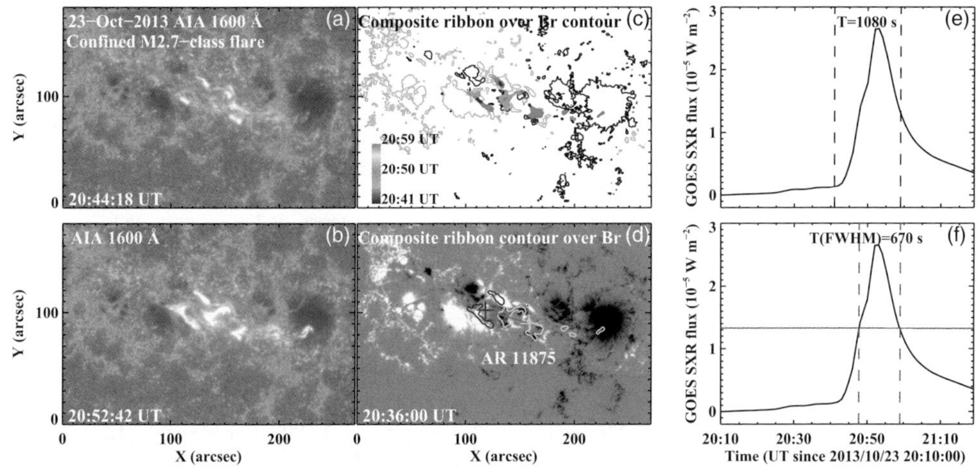

Figure 1. Left: snapshots of the 1600 Å flare ribbons in the M2.7-class confined flare in NOAA AR 11875 observed by AIA on 2013 October 23. Middle: composite ribbons plotted over the Br contours and the positive (blue) and negative (green) contours of composite ribbons over the Br magnetogram. Blue and green pluses respectively denote the positive and negative area-weighted centroids, and the red line in panel (d) represents the ribbon distance. Right: GOES 1-8 Å light curves of this event showing the flare start and end times (dashed lines in panel (e)), and the FWHM duration (panel (f)).

the confined and eruptive events are 1.5×10^{20} cm^2 and 1.2×10^{20} cm^2 (difference=22%), respectively. The normalized field strength of ARs for confined flares is slightly larger than that for eruptive events (panel (d1)). For ribbon parameters, the reconnection flux and ribbon distance are almost the same for eruptive and confined flares (panels (b2) and d(3)). Confined events have a slightly smaller ribbon area (differences of the log-averages of about 15% in panel (c2)) and have a slightly larger normalized field strength of the ribbon (differences of the log-averages of about 14% in panel (d2)).

In Figures 2(b3) and (c3), we display the histograms of flare reconnection flux ratio (Φ_{ribbon}/Φ_{AR}) and ribbon area ratio (S_{ribbon}/S_{AR}) for confined and eruptive events. It can be seen that the distributions of both flux ratio and area ratio show evident differences, with flux ratio and area ratio for confined events smaller than those for eruptive flares. Reconnection flux ratio ranges between 1% and 41% for eruptive flares and ranges between 1% and 21% for confined events. The log averages of flux ratio are 6.3% for confined and 9.5% for eruptive events. Similarly, the confined flares have the smaller area ratio (1%−18%) than eruptive events (1%−30%). The log-mean values of area ratio R_S are 4.0% for confined and 6.1% for eruptive cases.

Figure 3 shows 6 scatter plots of flare durations versus ribbon parameters for eruptive (blue circles) and confined (red circles) flares. It shows that the FWHM duration has a moderate correlation with ribbon distance, ribbon area and reconnection flux at the correlation coefficients CC of 0.51-0.53 (top in Figure 3). The decay duration also shows a moderate correlation with the two parameters of ribbon distance and reconnection flux at CC of 0.42-0.49 (bottom in Figure 3). The decay duration has a slightly stronger correlation with ribbon area at CC of 0.6 than FWHM duration. These relations are consistent with the results in Toriumi et al. (2017), who showed higher correlation coefficients among these parameters. We suggest that the difference in correlation coefficients is probably due to different ranges of flare classes in different statistical studies.

In our recent work, we extended our database to 722 flares ≥C5.0-class, including 255 eruptive flares and 467 confined events (Li et al. 2021). Figure 4(a) shows the scatter plot

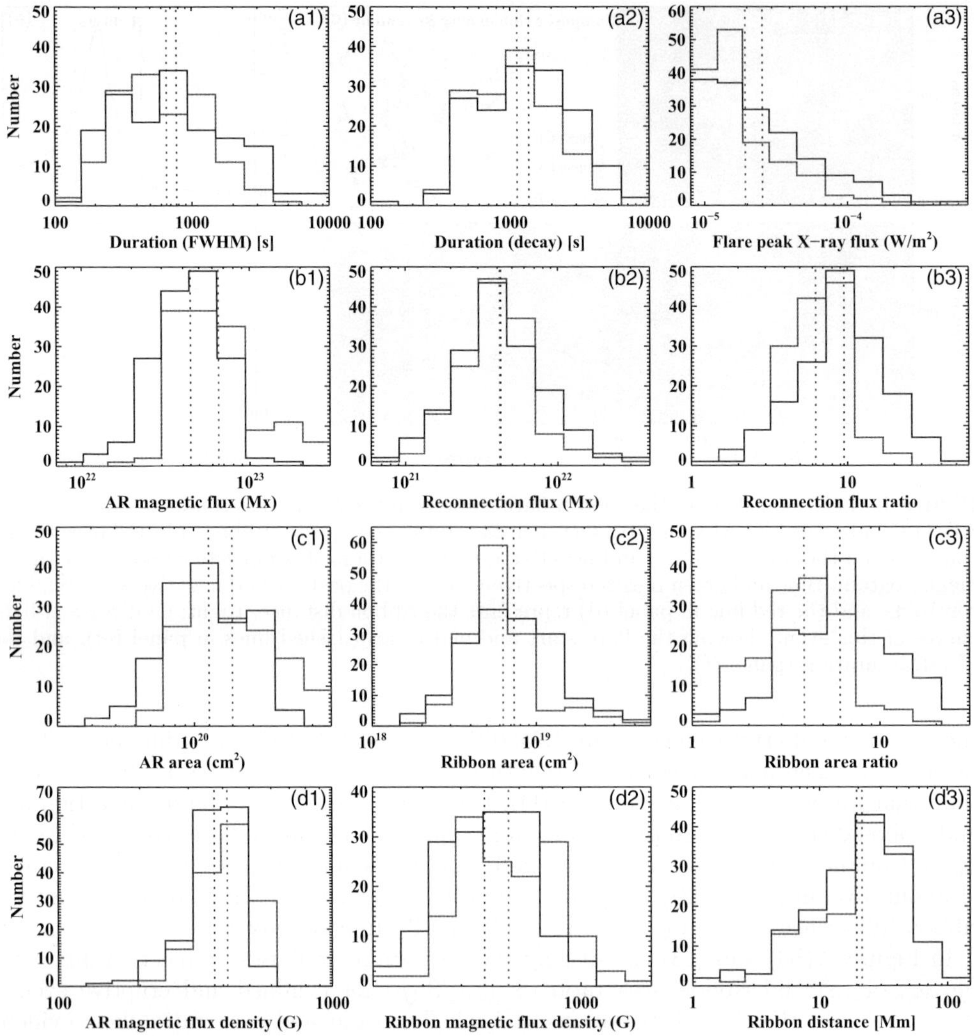

Figure 2. Comparison of the histograms for eruptive (blue) and confined (red) events: total event numbers are 170 and 152, respectively. Dashed vertical lines indicate the means of the log values. Top: histograms of FWHM duration, decay time, and GOES peak flux F_{SXR}. Second row: histograms of AR total magnetic flux Φ_{AR}, ribbon reconnection flux Φ_{ribbon}, and their ratio Φ_{ribbon}/Φ_{AR}. Third row: histograms of AR area S_{AR}, ribbon area S_{ribbon}, and their ratio S_{ribbon}/S_{AR}. Bottom: histograms of normalized field strength of ARs and ribbon areas, and ribbon distance.

of Φ_{AR} versus flare peak SXR flux F_{SXR}. Blue (red) circles are the eruptive (confined) flares. Obviously, when Φ_{AR} is large enough ($> 1.0\times10^{23}$ Mx; horizontal black dashed line), an overwhelming majority (about 97%) of flares do not generate CMEs (57 of 59 flares are confined). The value of Φ_{AR} for the 722 flares ranges from 8.5×10^{21} Mx to 2.3×10^{23} Mx, and we divide Φ_{AR} into 5 subintervals. Figures 4(b)-(f) shows the distributions of eruptive and confined flares in these 5 Φ_{AR} subintervals. It can be seen that in the smallest Φ_{AR} subinterval eruptive flares are not less than confined events, however, this reverses in the largest Φ_{AR} subinterval. Figure 5(a) shows the relations of the CME association rate R with F_{SXR} within the 5 Φ_{AR} subintervals. For each subinterval, R clearly increases with F_{SXR}, similar to previous results that studied the

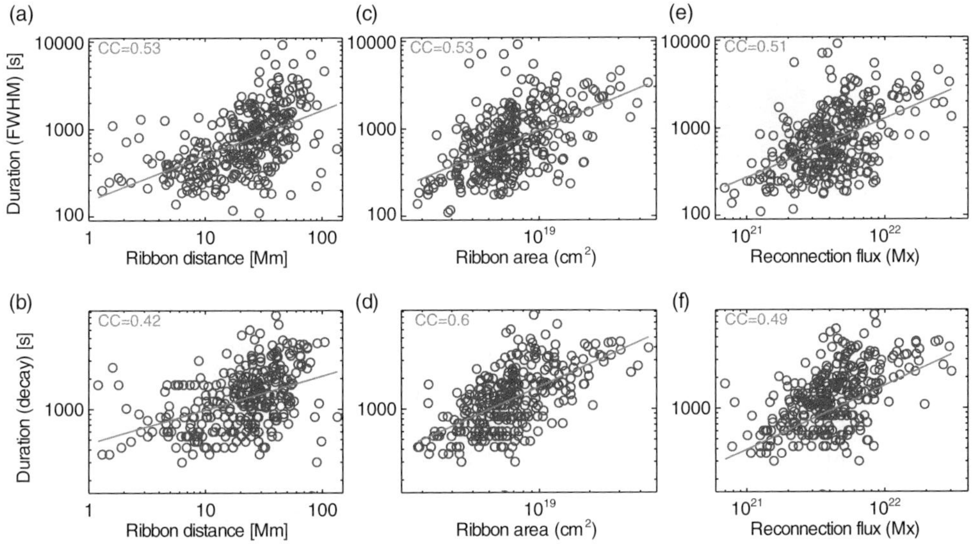

Figure 3. Scatter plots of flare durations versus ribbon parameters for eruptive (blue circles) and confined (red circles) flares. Orange straight line shows the result of a linear fitting to the log–log plots, while correlation coefficient CC is shown at the top left.

CME-flare association as function of flare class during the SOHO era (Yashiro et al. 2006). Each straight line in Figure 5(a) shows the linear fit

$$R = \alpha \log F_{SXR} + \beta, \tag{2.1}$$

where R is in percentage and F_{SXR} is in units of W m^{-2}. For the smallest Φ_{AR} subinterval ($\leq 2.0\times10^{22}$ Mx), the slope α is 113.8±13.1 and R reaches 100% when the flare is larger than M1.3 class (red straight line). For the subinterval of $2.0<\Phi_{AR}\leq3.5\times10^{22}$ Mx, the slope α decreases to 82.0±10.6 and R reaches 100% when the flare is larger than M2.4 class (green straight line). It can be seen that in ARs with a small Φ_{AR}, about 50% C5.0-class flares have associated CMEs. Only about 20% C5.0-class flares have associated CMEs in moderate Φ_{AR} subintervals. In ARs with the largest Φ_{AR} ($> 9.0\times10^{22}$ Mx), R decreases significantly when compared to subintervals characterized by smaller Φ_{AR} (black straight line). The relation between R and F_{SXR} in the largest Φ_{AR} subinterval is

$$R = (22.9 \pm 3.8) \log F_{SXR} + (125.7 \pm 17.9). \tag{2.2}$$

Based on the above equation, only 20% of all M-class flares originating from the largest ARs have associated CMEs and the rate R reaches about 40% for flares >X2. Almost all of C5.0-class flares are confined due to the strong constraining fields in the largest ARs.

Figure 5(b) shows the relation of the slope k with Φ_{AR}. Φ_{AR} is here defined as the mean of the individual log values in each Φ_{AR} subinterval. The plot shows that the slope k decreases monotonically with increasing Φ_{AR}. By assuming ARs in solar-type stars of $\Phi_{AR} \sim 1.0\times10^{24}$ Mx (Maehara et al. 2012; Shibata et al. 2013), and considering a value of the slope $k = 22.9$ in the largest ARs minus the average uncertainty estimates (~7.8, corresponding to the average error of five diamonds) as the slope k in solar-type stars, we estimate that the slope k might be no more than 15.1±7.8 (red circle). If C5.0-class flares are all confined on solar-type stars (similar to the subinterval of $\Phi_{AR} > 9.0\times10^{22}$ Mx), we can extrapolate the flare-CME association rate for solar-type stars is given as

$$R = (15.1 \pm 7.8) \log F_{SXR} + 80.0. \tag{2.3}$$

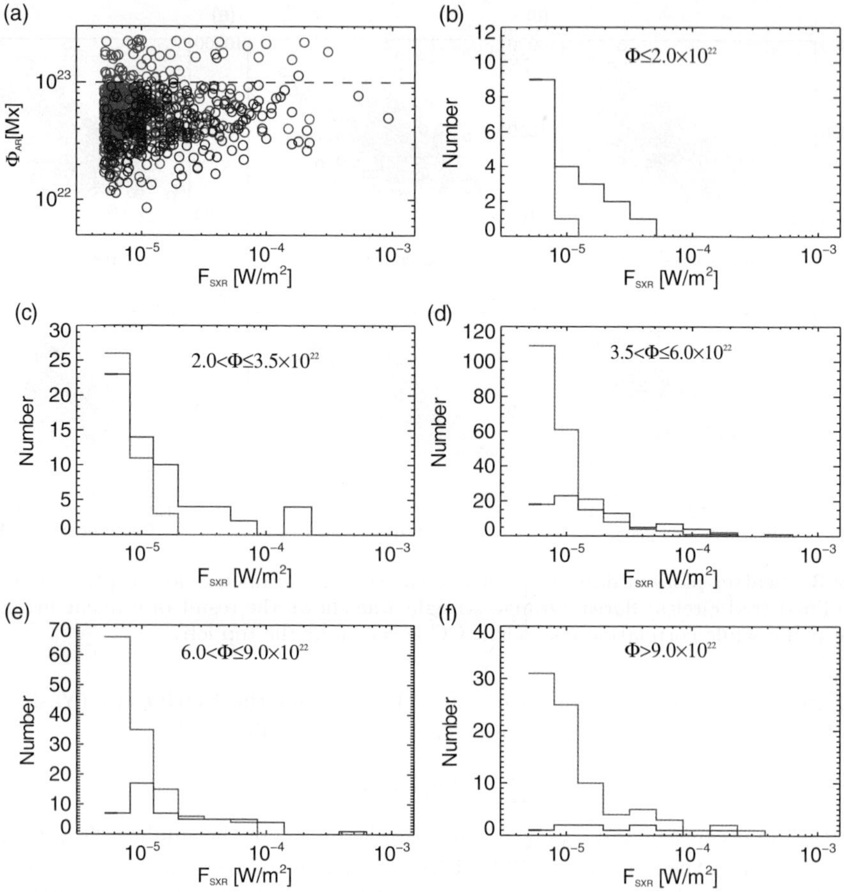

Figure 4. Panel (a): scatter plot of Φ_{AR} vs. F_{SXR}. Blue (red) circles are the eruptive (confined) flares ($\geq$ C5.0-class). Black dashed line corresponds to Φ_{AR} of 1.0×10^{23} Mx. Panels (b)-(f): distributions of eruptive and confined flares in 5 Φ_{AR} subintervals. Blue lines show eruptive flares and red lines show confined events.

Thus, for X100-class superflares in solar-type stars, the estimated association rate R is no more than 50%.

We also estimated the critical decay index heights h_{crit} for 82 events and showed the scatter plot of h_{crit} versus Φ_{AR} in Figure 6(a). It can be seen that h_{crit} increases with Φ_{AR}. This indicates that ARs with a larger magnetic flux tend to have stronger constraining field. The critical decay index height has a strong correlation with AR magnetic flux. The linear fitting to the scatter plot provides the relation of

$$h_{crit} = (38.31 \pm 2.37) \log \Phi_{AR} + (-834.53 \pm 53.92). \quad (2.4)$$

where h_{crit} and Φ_{AR} are in units of [Mm] and [Mx], respectively.

Using this equation, an Φ_{AR} value of 3.0×10^{22} Mx yields a h_{crit} of ~27 Mm (left vertical and bottom horizontal lines in Figure 6(a)) and 1.0×10^{23} Mx corresponds to h_{crit} of about 47 Mm (right vertical and top horizontal lines in Figure 6(a)). In Figure 6(b), we plot the flare peak X-ray flux versus h_{crit}. All flares (28 events) with a h_{crit} value smaller than 27 Mm are eruptive (Area A in Figure 6(b)), and about 95% (20 of 21) of events with h_{crit} larger than 47 Mm are confined (Area C in Figure 6(b)). These results suggest that stronger strapping fields over the ARs with a larger magnetic flux play the major role in confining the eruption.

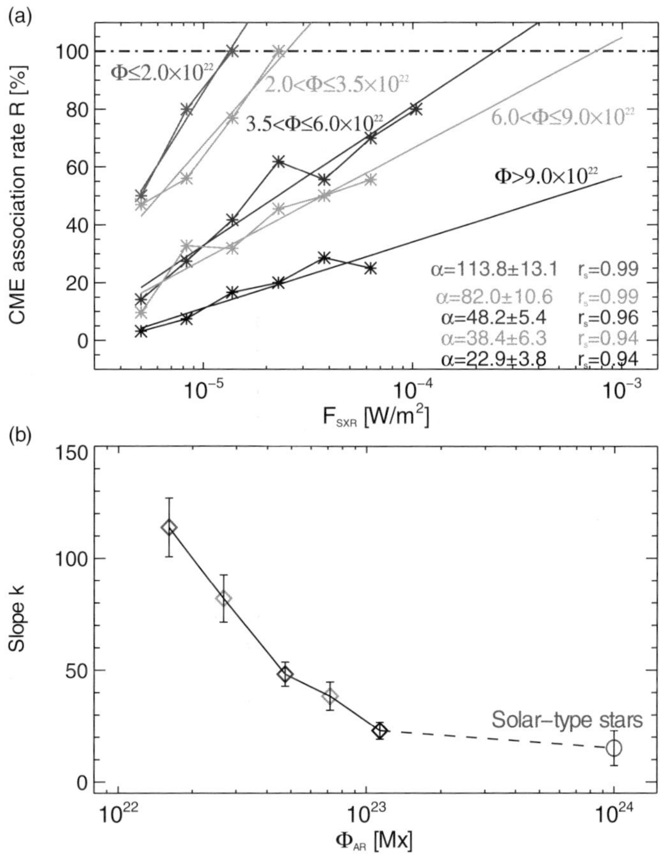

Figure 5. Panel (a): flare-CME association rate R as a function of F_{SXR} separately for 5 different Φ_{AR} subintervals. The colored straight lines show the results of linear fitting, and slopes α and Spearman rank order correlation coefficients r_s are shown at the bottom right. Panel (b): plot of slope k vs. Φ_{AR} (plotted at the average of the log values in each subinterval). Colored diamonds denote the slopes k in five different Φ_{AR} subintervals. The red circle is the estimated value of slope k (about 15.1) for solar-type stars by assuming Φ_{AR} of 1.0×10^{24} Mx (Maehara et al. 2012; Shibata et al. 2013).

We further calculated the magnetic non-potentiality parameters of ARs producing eruptive and confined flares. We use a subset of 106 flare events $\geq$M1.0 (43 eruptive and 63 confined) from a large database of 322 M-class flares during the period of June 2010 to June 2019 (see details in Li et al. 2022). The subset is selected based on the characteristics of ARs, and the selected ARs must fulfill the following two selection criteria. First, the ARs are flare-active and produced $\geq$3 M-class flares. Second, the ARs can be unambiguously classified into CME-active and CME-quiet, i.e., all the flares from the same AR have almost the identical eruptive character (ARs with only one exceptional flare event are also selected). A total of 21 ARs fulfill these selection criteria, including 12 CME-active ARs and 9 CME-quiet ARs. Among dozens of non-potential parameters, we select the twist parameter α of the magnetic field, which is the ratio between the vertical current density and the vertical magnetic flux density. For each event, we identified a "flaring polarity inversion line" (FPIL) mask to demarcate the core of an AR by using the method of Sun et al. (2015). We first find the polarity inversion line (PIL) pixels from a smoothed vertical magnetic field B_z, and dilate them with a circular kernel with a radius of 18 pixels (about 6.5 Mm; other radii are also used and the results are not affected).

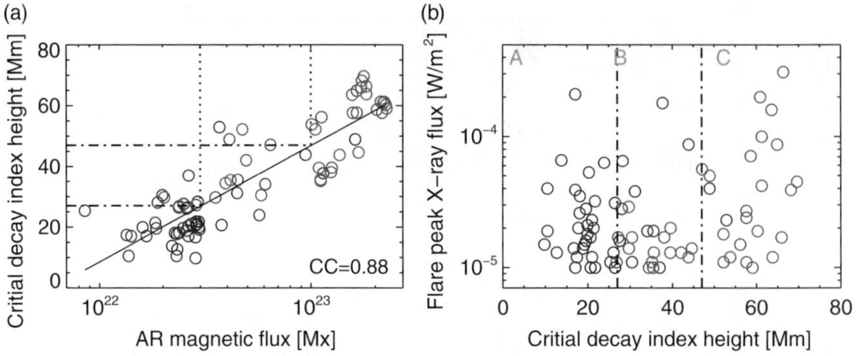

Figure 6. Scatter plots of critical decay index height vs. unsigned AR magnetic flux and flare peak X-ray flux vs. critical decay index height. Blue (red) circles are the eruptive (confined) flares. The black solid line in panel (a) shows the result of a linear fitting, and correlation coefficient CC is shown at the bottom right. Two vertical dotted lines in panel (a) denote the positions where Φ_{AR} respectively equals 3.0×10^{22} Mx and 1.0×10^{23} Mx. Two horizontal dash-dotted lines in panel (a) and two vertical dash-dotted lines in panel (b) respectively refer to critical decay index height of 27 Mm and 47 Mm.

Then we use the 1600 Å image near the flare peak from the AIA on board the *SDO*, and identify the flare ribbon areas by using a large kernel (above 700 DN s^{-1} which is about 10 times the standard deviation above the mean of the quiet-Sun values). Finally the intersection of dilated PIL and ribbon areas is considered as the FPIL mask. We calculate distributions of vertical electric current density J_z and the mean characteristic twist parameter $\alpha_{\rm FPIL}$ within the FPIL mask region.

Figure 7 shows four examples of two eruptive and two confined flares, and includes 1600 Å images (left), photospheric magnetograms (middle) and derived vertical electric current density J_z maps (right). We can see that for two eruptive flares (X2.1 flare in AR 11283 and M6.5 in AR 12371) the positive and negative currents have a coherent structure around the PIL (Figures 7(a)-(b)), indicating the presence of "current ribbons" as in a coherent flux rope. However, the two confined events (X3.1 in AR 12192 and M6.1 in AR 12222) exhibit disordered current distributions and do not have any noticeable structure (Figures 7(c)-(d)). It also can been seen that the FPIL mask regions (orange and black contours) overlap the area of large currents, implying that the FPIL mask corresponds to the AR core with the strongest magnetic non-potentiality.

Based on the derived J_z map, we then calculated the mean characteristic twist parameter $\alpha_{\rm FPIL}$ within the FPIL mask region for the 106 flares. Figure 8(a) shows the scatter plot of $\alpha_{\rm FPIL}$ versus $\Phi_{\rm AR}$. Blue (red) circles are the eruptive (confined) flares. It needs to be noted that $\alpha_{\rm FPIL}$ is a signed parameter and in our study $\alpha_{\rm FPIL}$ means its absolute value. Obviously, for $\alpha_{\rm FPIL}<0.07$ Mm^{-1} (black dotted line in Figure 8(a)), all the flares are confined (a total of 26 confined events). That is, all the eruptive flares have $\alpha_{\rm FPIL}\geq0.07$ Mm^{-1} (a total of 43 eruptive flares). For $\Phi_{\rm AR}<3.5\times10^{22}$ Mx (left green dashed line in Figure 8(a)), all the flares are eruptive. For $\Phi_{\rm AR}>1.0\times10^{23}$ Mx (right green dashed line in Figure 8(a)), all the flares are confined. Figure 8(b) shows the scatter plot of flare peak X-ray flux versus $\alpha_{\rm FPIL}/\Phi_{\rm AR}$. About 93% (40 of 43) of eruptive events have $\alpha_{\rm FPIL}/\Phi_{\rm AR}\geq2.2\times10^{-24}$ Mm^{-1} Mx^{-1}, and $\sim$ 83% (52 of 63) of confined flares have $\alpha_{\rm FPIL}/\Phi_{\rm AR}<2.2\times10^{-24}$ Mm^{-1} Mx^{-1} (black dash-dotted line in Figure 8(b)). This shows that the new relative non-potential parameter $\alpha_{\rm FPIL}/\Phi_{\rm AR}$ is well able to distinguish the two populations of eruptive and confined flares. However, there is still a small overlap (14 of 106) by using our criterion and the exceptional events are mainly from a CME-active AR 11302 and two CME-quiet ARs 11476 and 12268.

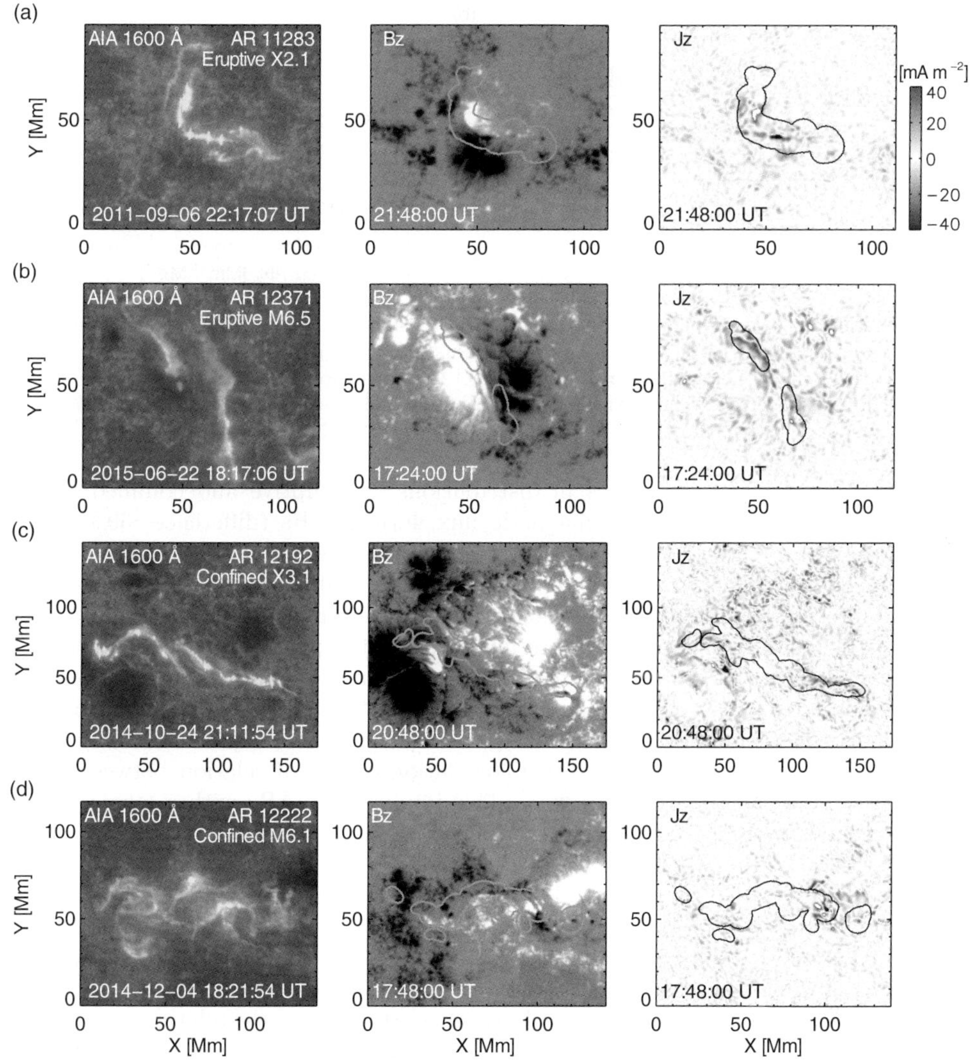

Figure 7. Four examples of two eruptive and two confined flares showing SDO/AIA 1600 Å images (left), SDO/HMI photospheric magnetograms B_z (middle) and derived vertical electric current density J_z maps (right). From top to bottom: eruptive X2.1-class flare in AR 11283, eruptive M6.5-class flare in AR 12371, confined X3.1-class flare in AR 12192 and confined M6.1-class flare in AR 12222. AIA 1600 Å images were remapped with CEA projection. Orange and black contours outline the FPIL mask regions within which the mean characteristic twist parameter (α_{FPIL}) in Figure 8 was calculated.

3. Conclusions

Based on the statistical analysis of AR magnetic parameters and ribbon parameters for eruptive and confined flares, our study delivered the following main results.

1. We find that there is no significant difference in distributions of durations and ribbon area between the eruptive and non-eruptive cases (differences of the log-averages of 15%-19%). Flare duration shows a moderate correlation with ribbon area and reconnection magnetic flux at correlation coefficients CC of about 0.5-0.6.

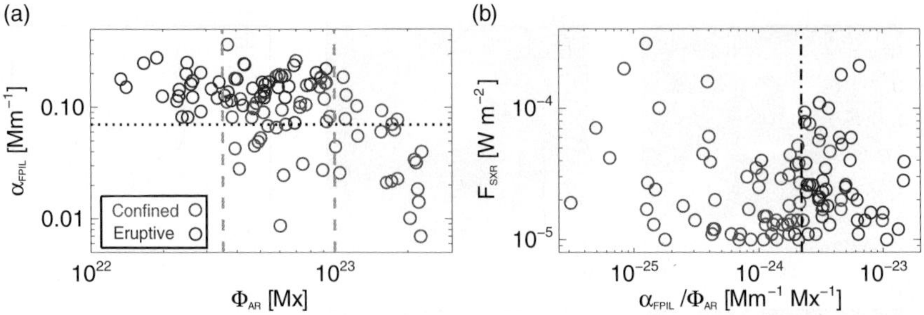

Figure 8. Scatter plots of mean characteristic twist parameter $\alpha_{\rm FPIL}$ versus total unsigned magnetic flux $\Phi_{\rm AR}$ and flare peak X-ray flux $\rm F_{SXR}$ versus $\alpha_{\rm FPIL}/\Phi_{\rm AR}$. Blue (red) circles are eruptive (confined) flares. Two vertical green lines in panel (a) correspond to $\Phi_{\rm AR}$ of 3.5×10^{22} Mx and 1.0×10^{23} Mx, respectively. The horizontal black line in panel (a) refers to $\alpha_{\rm FPIL}$ of 0.07 Mm^{-1}. The vertical black line in panel (b) denotes $\alpha_{\rm FPIL}/\Phi_{\rm AR}$ of 2.2×10^{-24} Mm^{-1} Mx^{-1}.

2. One of the clear differences in distributions for eruptive and confined flares is seen in the total unsigned magnetic flux Φ_{AR} of ARs (difference=39%). When Φ_{AR} is large enough ($> 1.0\times10^{23}$ Mx), an overwhelming majority (about 97%) of flares do not generate CMEs (57 of 59 flares are confined). We divided Φ_{AR} into 5 subintervals, and find that the flare-CME association rate R clearly increases with $\rm F_{\it SXR}$, i.e., larger flares are more likely associated with a CME. The slope of R depends on the total flux of the AR that produces the flare, and reveals a steep monotonic decrease with Φ_{AR}. This means that flares of the same GOES class but originating from an AR of larger Φ_{AR}, are much more likely confined.
3. We also find a very high positive correlation empirical relation between critical decay index height h_{crit} and Φ_{AR}. This implies that ARs with a large magnetic flux have a strong magnetic cage, which confines the eruption.
4. The ratio of the AR magnetic flux that is involved in the flare reconnection process ranges between 1% and 41% for eruptive flares and between 1% and 21% for confined events. Similarly, the confined flares have the smaller area ratio $\rm R_{\it S}$ (1%−18%) than eruptive events (1%−30%).
5. By analyzing the mean characteristic twist parameters α_{FPIL} within the "flaring polarity inversion line" region for 106 events, we find that the relative non-potential parameter $\alpha/\Phi_{\rm AR}$ could distinguish effectively between the eruptive from the non-eruptive cases. In about 90% of eruptive flares, $\alpha/\Phi_{\rm AR}$ is beyond the critical value (2.2×10^{-24} Mm^{-1} Mx^{-1}), whereas they are less than the critical value in $\sim$ 80% of confined flares.

In our work, we find that confined flares have similar distributions of duration, ribbon area and ribbon distance compared with eruptive flares. Harra et al. (2016) analyzed 42 X-class flares and found that some X-class eruptive flares showed a very impulsive rise time, and short durations, and some X-class confined flares were long-duration events. Toriumi et al. (2017) showed that there are no obvious differences in duration and ribbon area for eruptive and confined flares. As showed in the standard (CSHKP) flare model, the flare ribbons correspond to the footpoints of newly reconnected post-flare loops. The ribbon distance indicates the footpoint separation of the representative post-flare loop. If the loop configuration does not differ much for different flare events, the loop half length, L, would be proportional to ribbon distance. In the traditional viewpoint, eruptive flare have large-scale hot post-flare loops observed in SXR and are of long duration (e.g., tens of minutes to a few hours), whereas a confined flare shows brightening in compact loop structures that lasts for only a short period (e.g., less than an hour) (Kushwaha et al.

2014; Sahu et al. 2022). Our statistical results are not consistent with the traditional viewpoint, showing that confined flares are not compact and short-duration, and have similar lifetime and spatial scale with eruptive flares.

It was found that flare duration is correlated with magnetic reconnection flux and ribbon area. Toriumi et al. (2017) and Reep & Knizhnik (2019) have also investigated the relationship between the ribbon properties and flare duration. Toriumi et al. (2017) surveyed 51 flare $\geq$M5.0-class and found the correlation coefficients of flare duration versus reconnection flux and ribbon area are 0.6−0.8, larger than CC (0.5-0.6) in our study. Recently, Reep & Knizhnik (2019) analyzed almost 3000 flares and showed that the duration of GOES light curves is correlated with the magnetic reconnection flux and ribbon area in large flares at the Spearman rank order correlation coefficient r_s of 0.5 and 0.8 for M and X-class flares, respectively. They appear uncorrelated in smaller flares, which is likely due to the large errors in the measurements. It seemed easy to understand the relations: as more magnetic flux is involved, the reconnection processes continue longer.

Our results show that the magnetic flux of ARs is a key parameter in determining the eruptive character of large solar flares, and the flare-CME association rate is decreasing with the increasing magnetic flux of ARs. This finding is further supported by the high correlation obtained between Φ_{AR} and the critical height for torus instability. Φ_{AR} can be considered to be both a measure of the total flux that is in principle available for flaring as well as being a measure of the background field confinement overlying the flaring region. Our findings imply that the latter is the more important factor here. Large Φ_{AR} means a strong confinement and thus the flare-CME association rate is relatively low compared to small Φ_{AR}. Moreover, based on solar observations, we can speculate the associate rate R on solar-type stars by assuming Φ_{AR} of 1.0×10^{24} Mx (Maehara et al. 2012; Shibata et al. 2013). For X100-class "superflares" on solar-type stars, no more than 50% flares can generate stellar CMEs. Our findings provide an important contribution to revise the flare-CME association rates for solar-type stars, by including the distinct differences in these relations in dependence of the AR magnetic flux. This may provide an explanation why the detection of stellar CMEs is rare and the speeds of detected CME candidates are relatively low (e.g. Leitzinger et al. 2014; Veronig et al. 2021; Wu et al. 2022; Lu et al. 2022). Recently, Sun et al. (2022) used the solar magnetic field as a template to estimate the vertical extent of the torus-stable zone above a stellar active region, and suggested that the low apparent CME occurrence rate on cool stars is, at least partially, ascribed to the presence of extended torus-stable zone, which indicates the slowly decreasing confining background coronal magnetic field with height.

Moreover, it is also found that the flux ratio R_{flux} and area ratio R_S for confined flares are significantly smaller than those for eruptive events. This result is similar to the statistical result of Toriumi et al. (2017), who showed the parameter of the ribbon area normalized by the sunspot area determines whether a given flare is eruptive or not. They suggested that the relative structural relation between the flaring region and the entire AR controls the CME productivity.

By considering both the constraining effect of background magnetic fields and the magnetic non-potentiality of ARs, we propose a new parameter $\alpha/\Phi_{\mathrm{AR}}$ to describe the eruptive character of a flare, which is well able to distinguish flares associated with CMEs from flares that are not. Parameter α is the average characteristic twist of the magnetic field lines around the PILs of an AR, and provide measures of the nonpotentiality of AR core region (Leka & Skumanich 1999; Benson et al. 2021). Previous studies have shown that the parameter of magnetic twist plays an important role in discriminating between confined and eruptive events and can be used to predict whether an X- or M-class flaring AR would produce a CME (Bobra & Ilonidis 2016; Duan et al. 2019). Magnetic twist is thought to be related to the driving force of electric current-driven instabilities, such as the helical kink instability (Hood & Priest 1979). An AR containing a highly twisted magnetic field tends to produce an eruption when the twist exceeds a certain threshold. On the other hand, Φ_{AR} has a high positive correlation with the critical decay index height (related to the torus instability of a magnetic flux rope), implying that Φ_{AR} describes the strength of the background field confinement. Our statistical study reveals that the relative parameter α/Φ_{AR} has a better performance in distinguishing between the two types of flares than only Φ_{AR} or α does. We suggest that the relative parameter α/Φ_{AR} indicates the balance between the upward force that drives the eruptions and the downward force that suppresses the eruptions.

Acknowledgements

We are grateful to G. Cauzzi & A. Tritschler for organizing the Symposium IAUS 372. This work is supported by the B-type Strategic Priority Program of the Chinese Academy of Sciences (XDB41000000), the National Natural Science Foundations of China (12222306, 12273060) and Yunnan Academician Workstation of Wang Jingxiu (No. 202005AF150025).

References

Amari, T., Canou, A., Aly, J.-J., et al. 2018, Nature, 554, 211
Avallone, E. A. & Sun, X. 2020, ApJ, 893, 123
Baumgartner, C., Thalmann, J. K., & Veronig, A. M. 2018, ApJ, 853, 105
Benson, B., Pan, W. D., Prasad, A., et al. 2021, SoPh, 296, 163
Bobra, M. G. & Ilonidis, S. 2016, ApJ, 821, 127
Bobra, M. G., Sun, X., Hoeksema, J. T., et al. 2014, SoPh, 289, 3549
Duan, A., Jiang, C., He, W., et al. 2019, ApJ, 884, 73
Fletcher, L., Dennis, B. R., Hudson, H. S., et al. 2011, Space Science Reviews, 159, 19
Gupta, M., Thalmann, J. K., & Veronig, A. M. 2021, A&A, 653, A69
Harra, L. K., Schrijver, C. J., Janvier, M., et al. 2016, SoPh, 291, 1761
Hood, A. W. & Priest, E. R. 1979, SoPh, 64, 303
Kazachenko, M. D., Lynch, B. J., Welsch, B. T., et al. 2017, ApJ, 845, 49
Kazachenko, M. D., Lynch, B. J., Savcheva, A., et al. 2022, ApJ, 926, 56
Kushwaha, U., Joshi, B., Cho, K.-S., et al. 2014, ApJ, 791, 23
Lammer, H., Lichtenegger, H. I. M., Kulikov, Y. N., et al. 2007, Astrobiology, 7, 185
Leitzinger, M., Odert, P., Greimel, R., et al. 2014, MNRAS, 443, 898
Leka, K. D. & Skumanich, A. 1999, SoPh, 188, 3
Li, T., Hou, Y., Yang, S., et al. 2020, ApJ, 900, 128
Li, T., Chen, A., Hou, Y., et al. 2021, ApJ Letters, 917, L29
Li, T., Sun, X., Hou, Y., et al. 2022, ApJ Letters, 926, L14
Liu, L., Wang, Y., Wang, J., et al. 2016, ApJ, 826, 119
Lu, H.-. peng., Tian, H., Zhang, L.-. yun., et al. 2022, A&A, 663, A140
Maehara, H., Shibayama, T., Notsu, S., et al. 2012, Nature, 485, 478
Pesnell, W. D., Thompson, B. J., & Chamberlin, P. C. 2012, Sol. Phys., 275, 3
Reep, J. W. & Knizhnik, K. J. 2019, ApJ, 874, 157
Sahu, S., Joshi, B., Sterling, A. C., et al. 2022, ApJ, 930, 41
Shibata, K., Isobe, H., Hillier, A., et al. 2013, PASJ, 65, 49
Song, Y. & Tian, H. 2018, ApJ, 867, 159
Sun, X., Bobra, M. G., Hoeksema, J. T., et al. 2015, ApJ Letters, 804, L28
Sun, X., Török, T., & DeRosa, M. L. 2022, MNRAS, 509, 5075
Toriumi, S., Schrijver, C. J., Harra, L. K., et al. 2017, ApJ, 834, 56
Török, T. & Kliem, B. 2005, ApJ Letters, 630, L97
Vasantharaju, N., Vemareddy, P., Ravindra, B., et al. 2018, ApJ, 860, 58

Veronig, A. M., Odert, P., Leitzinger, M., et al. 2021, Nature Astronomy
Wang, D., Liu, R., Wang, Y., et al. 2017, ApJ Letters, 843, L9
Wang, Y., & Zhang, J. 2007, ApJ, 665, 1428
Wu, Y., Chen, H., Tian, H., et al. 2022, ApJ, 928, 180
Yashiro, S., Akiyama, S., Gopalswamy, N., et al. 2006, ApJ Letters, 650, L143

The Era of Multi-Messenger Solar Physics
Proceedings IAU Symposium No. 372, 2023
G. Cauzzi & A. Tritschler, eds.
doi:10.1017/S1743921323000285

Origin of extreme solar eruptive activity from the active region NOAA 12673 and the largest flare of solar cycle 24

Bhuwan Joshi and Prabir K. Mitra

Udaipur Solar Observatory, Physical Research Laboratory, Udaipur 313 001, India

Abstract. During 2017, when the Sun was moving toward the minimum phase of solar cycle 24, an exceptionally eruptive active region (AR) NOAA 12673 emerged on the Sun during August 28-September 10. During the highest activity level, the AR turned into a δ-type sunspot region, which manifests the most complex configuration of magnetic fields from the photosphere to the coronal heights. The AR 12673 produced four X-class and 27 M-class flares, along with numerous C-class flares, making it one of the most powerful ARs of solar cycle 24. Notably, it produced the largest flare of solar cycle 24, namely, the X9.3 event on 2017 September 6. In this work, we highlight the results of our comprehensive analysis involving multi-wavelength imaging and coronal magnetic field modeling to understand the evolution and eruptivity from AR 12673. We especially focus on the morphological, spectral and kinematical evolution of the two X-class flares on 6 September 2017. We explore various large- and small-scale magnetic field structures of the active region which are associated with the triggering and subsequent outbursts during the powerful solar transients.

Keywords. Sun: activity, Sun: corona, Sun: coronal mass ejections (CMEs), Sun: flares, Sun: magnetic fields.

1. Introduction

NOAA 12673 was one of the most flare productive solar ARs of cycle 24. It underwent a rapid evolution in magnetic complexity which resulted in tremendous flare productivity. Multi-instrument observations of NOAA 12673, coupled with modeling analysis of coronal magnetic fields, have provided us with a unique opportunity to investigate various physical parameters that play vital role in the onset of large flares and influence the early CME dynamics. Notably, NOAA 12673 produced the largest flare of the cycle, an X9.3 flare on 6 September 2017.

In view of its uniqueness in the flare productivity, naturally the active region has already been subjected to a number of studies that involve different aspects of the long-term evolution of the AR as well as flare case studies (e.g., Yang et al. 2017; Gary et al. 2018; Hou et al. 2018; Liu et al. 2019; Mitra et al. 2018, 2020b; Romano et al. 2018; Seaton & Darnel 2018; Verma 2018; Veronig et al. 2018; Moraitis et al. 2019; Chen et al. 2020, etc.).

In this article, we investigate the evolution and magnetic characteristics of NOAA 12673. We briefly present the multi-wavelength imaging of the homologous X-class flares of intensities X2.2 and X9.3 occurred on 6 September 2017 to understand the triggering mechanism and stages of energy release. The complexity of photospheric and coronal magnetic fields prior to the X-class flares are explored to assess the link between the large- and small-scale magnetic field structures of the solar active region.

Table 1. Summary of the large flares (GOES class M1 or above) originated from the active region NOAA 12673.

Flare Id.	**Date (Sep '17)**	**AR conf.**	**Location**	**Flare timing (UT) Start/**	**Peak/**	**End**	**GOES class**
F_1	4	$\beta\gamma$	S10W04	05:36/	05:49/	06:05	M1.2
F_2			S10W08	15:11/	15:30/	15:33	M1.5
F_3			S07W11	18:05/	18:22/	18:31	M1.0
F_4			S09W11	18:46/	19:37/	19:52	M1.7
F_5			S10W11	19:59/	20:02/	20:06	M1.5
F_6			S10W11	20:28/	20:33/	20:37	M5.5
F_7			S09W12	22:10/	22:14/	22:19	M1.7
F_8	5	$\beta\gamma\delta$	S09W14	01:03/	01:08/	01:11	M4.2
F_9			S09W15	03:42/	03:51/	04:04	M1.0
F_{10}			S11W18	04:33/	04:53/	05:07	M3.2
F_{11}			S11W19	06:33/	06:40/	06:43	M3.8
F_{12}			S10W23	17:37/	17:43/	17:51	M1.5
F_{13}	6	$\beta\gamma\delta$	S08W32	08:57/	09:10/	09:17	X2.2
F_{14}			S09W34	11:53/	12:02/	12:10	X9.3
F_{15}			S08W36	15:51/	15:56/	16:03	M2.5
F_{16}			S08W38	19:21/	19:30/	19:35	M1.4
F_{17}			S08W40	23:33/	23:39/	23:44	M1.2
F_{18}	7	$\beta\gamma\delta$	S08W44	04:59/	05:02/	05:08	M2.4
F_{19}			S07W46	09:49/	09:54/	09:58	M1.4
F_{20}			S07W46	10:11/	10:15/	10:18	M7.3
F_{21}			S08W48	14:20/	14:36/	14:55	X1.3
F_{22}			S11W54	23:50/	23:59/	00:14 (8 Sep)	M3.9
F_{23}	8	$\beta\gamma\delta$	S09W55	02:19/	02:24/	02:29	M1.3
F_{24}			S07W55	03:39/	03:43/	03:45	M1.2
F_{25}			S09W57	07:40/	07:49/	07:58	M8.1
F_{26}			S09W63	15:09/	15:47/	16:04	M2.9
F_{27}			S08W69	23:33/	23:44/	23:56	M2.1
F_{28}	9	$\beta\gamma\delta$	S11W70	04:14/	04:28/	04:43	M1.1
F_{29}			S08W74	11:50/	11:04/	11:42	M3.7
F_{30}			S09W88	22:04/	23:53/	01:30 (10 Sep)	M1.2
F_{31}	10	$\beta\gamma\delta$	S08W88	15:35/	16:06/	16:31	X8.2

2. NOAA 12673: Morphology, evolution, and eruptivity

2.1. *Evolution and activity*

The active region NOAA 12673 appeared on the eastern limb of the Sun as a simple unipolar sunspot. Subsequently, the active region exhibited the constant emergence of magnetic flux of both the polarities as well as photospheric motions including shearing and rotational motions (Figure 1). This led to the quick evolution of the region into the most complex configuration, namely the δ-type, resulting a series of large eruptive flares. Notably, the AR produced 4 X-class and 27 M-class flares along with numerous smaller events. In Table 1, a summary of the flares of GOES class M and above is given. From Table 1, we note that the first M class flare from NOAA 12673 originated on 4 September 2017 following which 12 further M-class flares occurred within 2 days. After producing 2 homologous X-class flares on 6 September 2017 (see Section 2.2), the AR produced another X-class flare (X1.3) on the next day. Subsequently, on 10 September, a large X8.2 flare occurred while the AR was transiting toward the far side of the Sun.

2.2. *Homologous X-class flares on 6 September 2017*

The first X-class flare from NOAA 12673 (GOES class X2.2) originated on 6 September and within ≈3 hours after its initiation, the AR produced the largest flare of the solar cycle 24, having GOES class X9.3. Figure 2 represents a few representative 94 Å and 304 Å images showing the evolution of X2.2 flare. The EUV images prior to the onset of

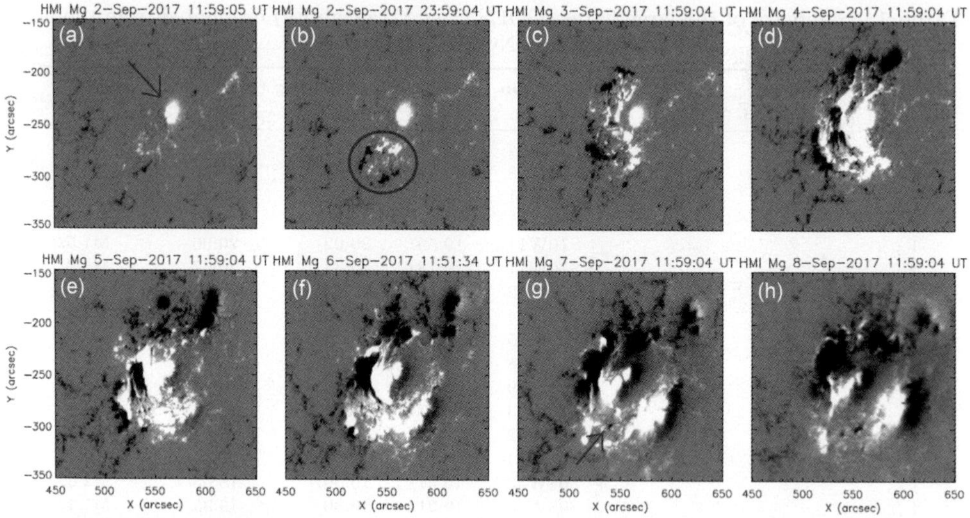

Figure 1. Series of HMI continuum images showing the build up and evolution of the active region NOAA 12673. All the images were rotated to the co-ordinate corresponding to 6 September 2017 12:00 UT. The arrow in panel (a) indicate the positive polarity sunspot. The oval shape in panel (b) indicates the initial flux emergence on 3 September 2017. The arrow in panel (g) indicates emergence of negative flux regions within the positive polarity regions during the decay phase of the active region.

the flare (Figures 2(a) and (d)) readily confirms that the coronal configuration associated with the AR took the shape of an impressive 'inverse-S' i.e., a prominent coronal sigmoid (the sigmoidal structure is highlighted by the dotted curve in Figure 2(a)). Notably, the sigmoid encompassed the entire stretch of the AR and the axis of the sigmoid was associated with the PIL between the positive and negative flux regions. The intense emission observed in the 94 Å images along the coronal sigmoid is suggestive of the presence of a hot corona channel i.e., an activated flux rope. During the impulsive phase of the flare, this flux rope underwent partial eruption with the formation of chromospheric flare ribbons.

After the X2.2 flare, the sigmoidal structure underwent a gradual expansion and rise in the intensity of emission (cf. Figures 2c and 3a). Notably, the X9.3 flare initiated with a localized spot-like brightening at the northern footpoint of the hot channel (indicated by the white arrow in Figure 3(b)). Almost simultaneously we observed prominent signatures of filament eruption which initially ascended with the hot channel (indicated by the arrow in the inset of Figure 3(f)). Following the eruption, we observed formation of two prominent flare ribbons which were situated close to each other (shown within the inset of Figure 3(g). The narrow post-reconnection arcade connecting these flare ribbons is indicated by the yellow arrow in Figure 3(c). Importantly, the gradual phase of the flare was characterized by the formation of a large-scale loop arcade and flare ribbons that encompassed the entire stretch of the AR (indicated by the blue arrows in Figures 3(d) and (h)). A set of extended flare ribbons at the footpoints of the large-scale loop arcade was identified in the AIA 304 Å images (indicated by the green arrows in Figure 3(h)).

2.3. *Complexity and topology of magnetic fields*

To understand the magnetic nature of the δ-spots of NOAA 12673 prior to X-class flares, in Figure 4, we plot co-temporal white light image and magnetogram of the AR during its most flaring phase and calculate magnetic gradient across the PIL associated

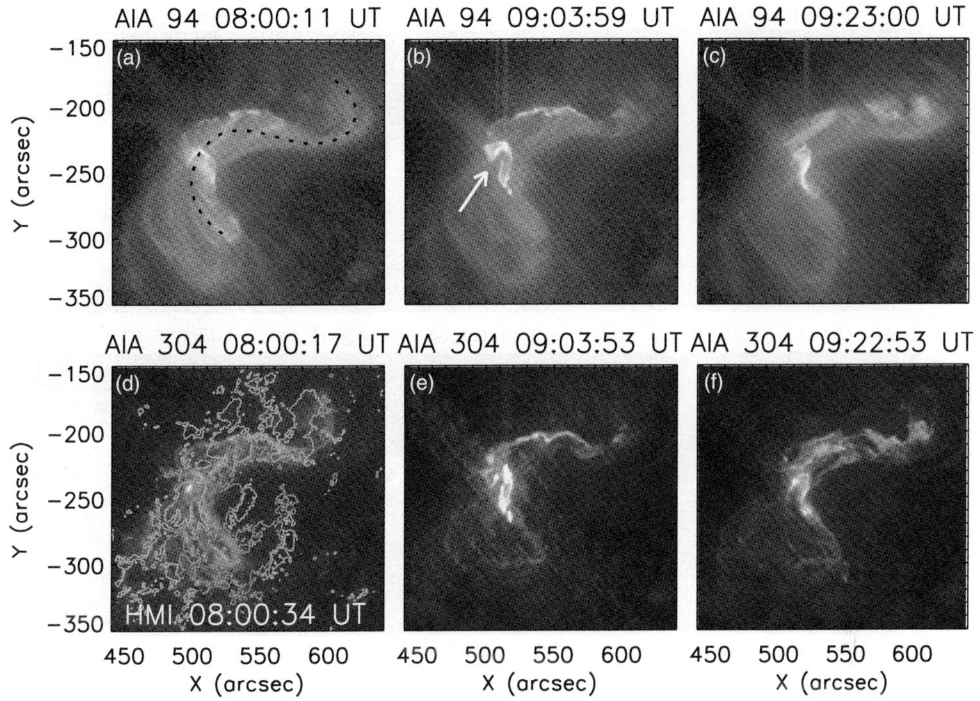

Figure 2. AIA 94 Å (panels (a)-(c)) and 304 Å (panels (d)-(f)) showing the evolution of the X2.2 flare on 6 September 2017. The dashed curve in panel (a) highlights the pre-flare sigmoidal structure. The arrow in panel (b) indicates a hot channel. Co-temporal HMI LOS magnetogram contours are plotted over panel (d). The green and sky-blue contours refer to positive and negative field, respectively. Contour levels are ±150 G.

with the bipolar region. The white light image suggests the presence of a few segmented umbrae within the central sunspot area which were surrounded by a single penumbra (Figure 4(a) and (b)). Comparison of the white light image and the magnetogram suggests a few of these fragmented umbrae to have negative polarity and a few to be associated with positive polarity. Our analysis readily suggests that LOS magnetic strength sharply changed from ≈1500 G to ≈-1000 G within a distance of ≈1 arcsec across the PIL. Consequently, the PIL region was associated with the highest magnetic gradient with a value $\approx 2.4\times10^3$ G Mm^{-1}. This confirms NOAA 12673 to be associated with all the characteristic properties of typical δ-type sunspots prior to the onset of the X-class flares.

In order to understand the coronal magnetic configuration leading to the onset of the two X-class flares, we carried out NLFFF extrapolation at 08:48 UT and 11:48 UT (prior to the onset of the X2.2 and X9.3 flares, respectively). The modeled magnetic field configuration prior to the onset of the X2.2 flare (Figures 5(a) and (b)) readily indicates the presence of a flux rope extending over the PIL (shown by the green lines) with northern and southern footpoints fixed at negative and positive polarity regions of the central sunspot region, respectively. Notably, the appearance of the hot channel, observed just prior to the onset of the X2.2 flare, was almost precisely the same as the shape of the flux rope (see Figure 2(b)). The location next to the northern footpoints of the flux rope was characterized by a complex coronal structure similar to the fan-spine configuration that involved a magnetic null point. The two sets of fan-spine lines are depicted by the blue and pink lines in Figures 5(a) and (b) and the coronal null point is indicated by a black arrow in Figure 5(b).

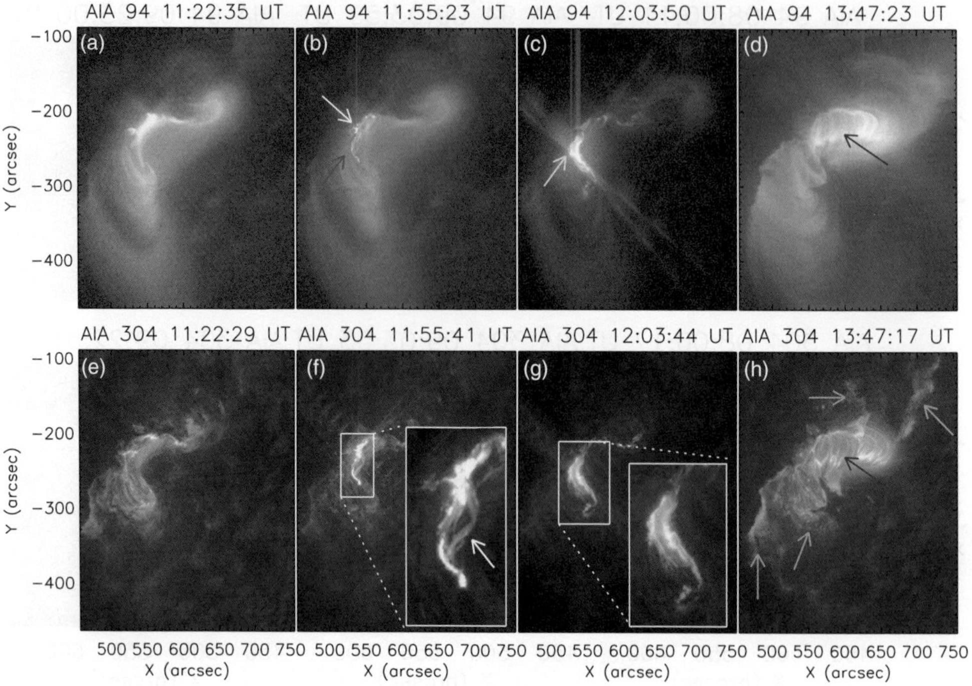

Figure 3. AIA 94 Å (panels (a)-(d)) and 304 Å (panels (e)-(h)) images showing the evolution of X9.3 flare on 6 September 2017. The red arrow in panel (b) indicates a hot channel. The white arrow in panel (b) indicates the location of the earliest brightening prior to the eruption of the hot channel. The yellow arrow in panel (c) indicates intense thermal emission from the location of the hot channel. The blue arrows in panels (d) and (h) indicate a large-scale post-reconnection arcade encompassing almost the entire span of the active region. The white arrow in the inset of panel (f) indicates the erupting hot filament. The green arrows in panel (h) indicate a set of large-scale flare ribbons during the gradual phase of the X9.3 flare.

The overall model magnetic configuration prior to the onset of the X9.3 flare at the core region of NOAA 12673 (Figures 5(c)-(d)) was similar to that corresponding to the X2.2 flare. However, the flux rope prior to the X9.3 flare (shown by the green lines in Figures 5(c)-(d)) appeared to be magnetically more twisted than that of the X2.2 flare. The overall fan-spine configuration prior to the X9.3 flare is shown by the blue and yellow lines in Figures 5(c) and (d). Notably, the northern footpoint of the flux rope itself appeared to be involved in the fan-spine configuration. In Figure 5(d), the null point involved in the fan-spine configuration is indicated by a black arrow. Here we remember that, the X9.3 initiated with a point-like brightening from the northern footpoint location of the hot channel (Figure 5(b)).

3. Conclusions

The AR NOAA 12673 was highly flare productive. Interestingly, the AR emerged on the visible part of the Sun with the simplest magnetic category of solar active regions (α-type). However, constant emergence of magnetic flux in the vicinity of the initial sunspot (Figure 1), coupled with photospheric motions enabled the active region to stand out from other active regions by making it one of the most complex active regions with extremely high flaring activity. It is well understood that photospheric motions lead to the storage of non-potential magnetic energy in the active regions making them flare productive (Aulanier 2014; Mitra et al. 2020a). Thus, despite having a simplistic initial

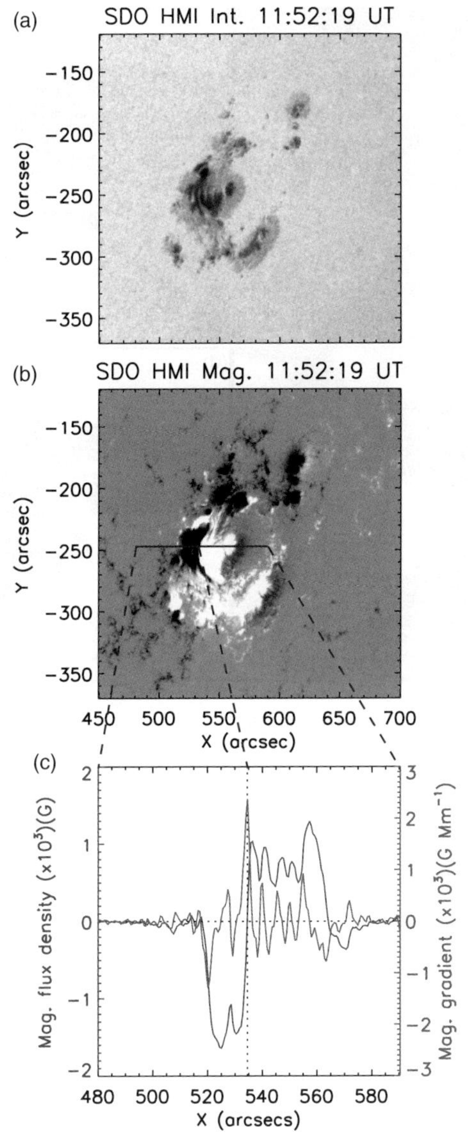

Figure 4. Panel (a): HMI continuum image of the active region NOAA 12763 prior to the onset of the X9.3 flare on 6 September 2017, showing multiple, fragmented umbrae within a single penumbrae. Panel (b): co-temporal HMI LOS magnetogram of NOAA 12673. Panel (c): LOS magnetic flux density (red) and magnetic gradient (green) computed along the blue slit in panel (b).

phase, the active region NOAA 12673 became capable of producing a series of large flares within a few days, thanks to the consistent flux emergence and continuous storage of non-potential energy (Mitra et al. 2018, 2020b).

During the occurrence of X-class flares, the complex AR had shown δ-sunspots, which are identified with a complex distribution of sunspot groups in which the umbrae of positive and negative polarities share a common penumbra. Such complex ARs are known to produce powerful flares (Takizawa & Kitai 2015). The δ-sunspots were concentrated at the central part of the AR where magnetic fields were very strong, and the magnetic field gradient across the PIL was extremely high. The reported eruptive activities in AR

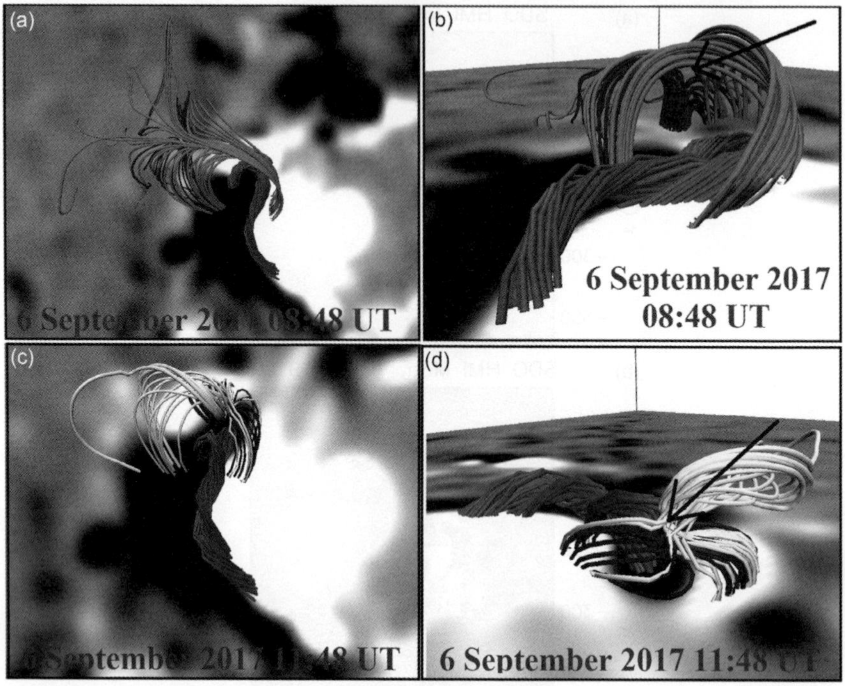

Figure 5. Reconstructed coronal magnetic field configuration by a NLFFF model (Wiegelmann et al. 2012) prior to the X2.2 flare (top panels) and X9.3 flare (bottom panels). In all the panels, the green lines represent magnetic flux ropes. The arrows in panels (b) and (d) indicate coronal null points.

12673, thus, represent, the capability of the AR in the rapid generation and storage of huge amount of excess magnetic energy in the corona. The evolution of free magnetic energy by Mitra et al. (2018) suggest that a large amount of free magnetic energy was already stored in NOAA 12673 before the flaring activities and that the large X-class flares of 6 September 2017 essentially released only a small fraction of it.

Our observations reveal a sigmoid structure (i.e., large, bright AR loops presenting S-shaped morphology; see Joshi et al. 2017) before the first X-class flare that remained intact after the event (Figure 2). However, the sigmoidal region completely transformed into the postflare arcade in a sequential manner during the second X-class flare (Figures 3). The present observations thus indicate the occurrence of two eruptive X-class flares within a single sigmoid-to-arcade event, which does not seem to be a commonly observed phenomenon. It is noteworthy that the flare ribbons corresponding to the X-class flares were subjected to very little separation, which is rather uncommon in strong eruptive X-class flares. We attribute this small separation between flare ribbons to the fact that footpoints of the flare loops were rooted in δ-sunspot regions of strong magnetic field **B**. Thus, although the loop growth and inner ribbon expansion speed **v** were small, the associated local reconnection rate would have been large enough to produce X-class events.

Both the X-class events not only occurred at the same region but also shared a common initiation process. The trigger happened at the central part of the sigmoid that spatially lay close to the northern edge of the high magnetic gradient area. The detection of preexisting magnetic null in NLFFF extrapolations, near the northern edge of the flux rope structure, has important implications in understanding the triggering mechanism for the subsequent eruptive events (see e.g., Prasad et al. 2020). The synthesis of observed and modeled coronal loops/field lines provides concrete evidence that the location of the initial energy release is spatially correlated with the site of the magnetic null. Our observations indicate that the triggering reconnections at the magnetic null were capable of destabilizing the flux rope only locally during the first X-class flare because the eruption was relatively minor and the overall sigmoid structure was preserved. This could happen because of the strong field lines lying over the flux rope, as well as firm tying of its southern footpoint. The coronal conditions seem to become conductive for the complete eruption by the time of the second X-class flare, which resulted in the eruptive phenomena at much larger scales.

In summary, flux emergence and photospheric motions stored large amount of non-potential magnetic energy in the active region NOAA 12673. The implications of the non-potential magnetic energy storage can be interpreted in the forms of the complex coronal structures found in the active region i.e., coronal sigmoid, magnetic flux rope, coronal null point etc. Thus, NOAA 12763 can be considered as a test-subject to examine the role of photospheric magnetic activities toward the triggering of powerful solar eruptions.

Acknowledgements

BJ acknowledges the IAU for providing the travel grant to attend the General Assembly.

References

Aulanier, G. 2014, in Nature of Prominences and their Role in Space Weather, ed. B. Schmieder, J.-M. Malherbe, & S. T. Wu, Vol. 300, 184–196
Chen, B., Yu, S., Reeves, K. K., & Gary, D. E. 2020, ApJ Letters, 895, L50
Gary, D. E., Chen, B., Dennis, B. R., et al. 2018, ApJ, 863, 83
Hou, Y. J., Zhang, J., Li, T., Yang, S. H., & Li, X. H. 2018, A&A, 619, A100
Joshi, B., Kushwaha, U., Veronig, A. M., et al. 2017, ApJ, 834, 42
Liu, L., Cheng, X., Wang, Y., & Zhou, Z. 2019, ApJ, 884, 45
Mitra, P. K., Joshi, B., & Prasad, A. 2020a, SoPh, 295, 29
Mitra, P. K., Joshi, B., Prasad, A., Veronig, A. M., & Bhattacharyya, R. 2018, ApJ, 869, 69
Mitra, P. K., Joshi, B., Veronig, A. M., et al. 2020b, ApJ, 900, 23
Moraitis, K., Sun, X., Pariat, É., & Linan, L. 2019, A&A, 628, A50
Prasad, A., Dissauer, K., Hu, Q., et al. 2020, ApJ, 903, 129
Romano, P., Elmhamdi, A., Falco, M., et al. 2018, ApJ Letters, 852, L10
Seaton, D. B. & Darnel, J. M. 2018, ApJ Letters, 852, L9
Takizawa, K. & Kitai, R. 2015, SoPh, 290, 2093
Verma, M. 2018, A&A, 612, A101
Veronig, A. M., Podladchikova, T., Dissauer, K., et al. 2018, ApJ, 868, 107
Wiegelmann, T., Thalmann, J. K., Inhester, B., et al. 2012, SoPh, 281, 37
Yang, S., Zhang, J., Zhu, X., & Song, Q. 2017, ApJ Letters, 849, L21

The Era of Multi-Messenger Solar Physics
Proceedings IAU Symposium No. 372, 2023
G. Cauzzi & A. Tritschler, eds.
doi:10.1017/S1743921323000236

Solar cycles reconstructed over the last millennium: Do Waldmeier and Gnevysev-Ohl rules work?

Markus Similä and Ilya Usoskin

Space Physics and Astronomy Research Unit and Sodankylä Geophysical Observatory, University of Oulu, Oulu, Finland

Abstract. Using tree-ring radiocarbon ^{14}C data, solar cycles are now reconstructed for the last millennium, more than doubling the previously known statistic of direct solar observations and giving a new opportunity to validate basic empirical rules connecting solar cycle lengths, strengths and intensities. This includes the Waldmeier rule relating the cycle's strength to the length of its ascending phase, and the Gnevyshev-Ohl rule suggesting that cycles are paired so that the intensity of an odd solar cycle is higher than that of the preceding even cycle. Using the extended solar-cycle statistic, we found that the Waldmeier rule remains robust for the well-defined solar cycles implying that it is an intrinsic feature of the solar cycle. However, the validity of the Gnevyshev-Ohl rule is not confirmed at any reasonable statistical level, indicating that either the insufficient accuracy of the reconstructed solar cycles or that this rule is not a robust feature.

Keywords. Sunspot numbers, Solar cycle, Waldmeier rule, Gnevyshev-Ohl rule

1. Introduction

The Sun is an active star whose magnetic activity is dominated by quasi-periodic 11-year cyclicity (Hathaway 2015) which is a result of the action of the solar dynamo in the convection zone (Charbonneau 2020). However, neither period/length/shape nor the amplitude of solar cycles is constant as cycles vary in time ranging from Grand minima of solar activity when hardly any spots appear on the Sun to Grand maxima of very high activity cycles Usoskin (2017). Solar activity is traditionally studied using data from direct telescopic solar observations started in 1610. This sunspot-number series covers 36/37 solar cycles of which 24 are well defined, 8 are poorly resolved and four (during the Maunder minimum) are unresolved. One cycle might have been lost ca. 1800 due to sparse data (Usoskin et al. 2009). Well-defined cycles start in the middle of the 18th century and are standardly represented by the international sunspot number series (ISN) version 2.0 (Clette & Lefèvre 2016) available at the SILSO database (`https://www.sidc.be/silso/datafiles`).

There are several rules which relate different parameters of the solar cycles to each other and across the cycles. These rules were established empirically and form observational constraints on the solar dynamo theory (e.g., Hathaway 2015; Charbonneau 2020).

The most famous rule is the Waldmeier rule (Waldmeier 1935, 1939) which says, in its classical definition, that strong cycles rise fast, viz. the duration of the ascending phase of a solar cycle is inversely related to its strength. The Waldmeier rule is fairly stable across the well-established part of the sunspot-number record (Usoskin et al. 2021a). Sometimes the Waldmeier rule is incorrectly interpreted as an inverse relation between

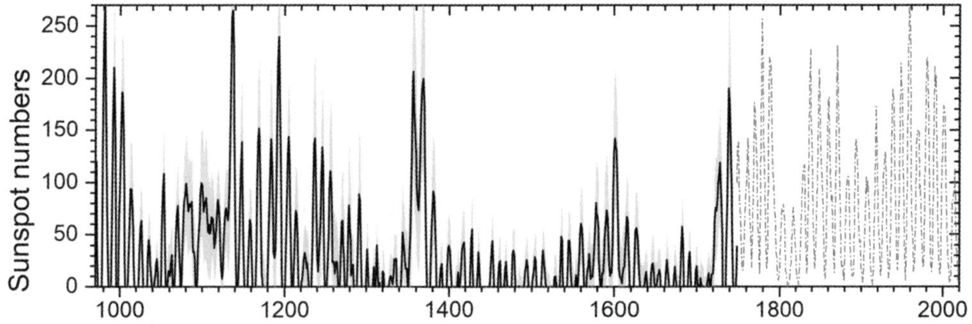

Figure 1. Annual sunspot numbers since 950 AD reconstructed from ^{14}C measurements (black curve) along with the 68% confidence intervals (Brehm et al. 2021; Usoskin et al. 2021b). ISN v.2.0 is shown with the red dashed curve (Clette & Lefèvre 2016).

the total length of a solar cycle and its amplitude. Here we use the classical definition of the Waldmeier rule.

Another empirical relation is the Gnevyshev-Ohl rule (Gnevyshev & Ohl 1948), which says that solar cycles are tied in even-odd pairs so that the integral intensity (the sum of sunspot numbers over a cycle) of the leading even-numbered cycle tends to be smaller than the intensity of the subsequent odd-numbered cycle. Sometimes it is called the even-odd effect in the literature. There are several potential breaks of the Gnevyshev-Ohl rule: one was around the Dalton minimum ca. 1800 and may be related to a lost solar cycle (e.g., Usoskin et al. 2001, 2009; Karoff et al. 2015); and another during the last few solar cycles as related to the end of the Modern grand maximum.

These empirical rules were in focus of many earlier studies (e.g., Usoskin et al. 2001; Hathaway et al. 2002; Solanki et al. 2002; Aparicio et al. 2012; Carrasco et al. 2016; Takalo & Mursula 2018) using different approaches and different versions of the sunspot series. Not much can be added to these rigorous analyses based on the sunspot number records for the period of about 300 years after the Maunder minimum.

However, a new opportunity to verify the validity of these empirical rules has appeared recently. A new dataset of radiocarbon Δ^{14}C in tree rings was conducted by Brehm et al. (2021) who used the acceleration mass spectrometry at ETH Zürich to measure radiocarbon with annual resolution since 950 AD with the high accuracy of 1.8 permil. Although the uncertainties are comparable with the amplitude of the 11-year cycle in Δ^{14}C that is greatly attenuated by the carbon cycle (e.g., Bard et al. 1997), the combination of the high cadence and accuracy made it possible to reconstruct the solar cycle over the last millennium (Usoskin et al. 2021b). The new reconstruction (shown in Figure 1 as the black curve with the 68% confidence interval) covers the period of 970 – 1900 AD and significantly enlarges the statistic of known solar cycles: 96 solar cycles are identified. The last millennium was a special period including a cluster of four Grand minima of solar activity – Oort, Wolf, Spörer and Maunder minima in the 11th, 14th, 15th and 17th centuries, respectively (Usoskin et al. 2007; Inceoglu et al. 2015). Solar cycles cannot be reliably defined during the Grand minima, leaving about 50 well and reasonably-well-defined cycles for the analysis. A preliminary analysis of the Waldmeier rule was performed by Usoskin et al. (2021b) suggesting that it remains statistically significant on the millennial time scale. However, as reported by Usoskin et al. (2022), solar cycles provided by Usoskin et al. (2021b) had erroneously large error bars that could affect the statistical study of the Waldmeier rule.

Here we report the results of a thorough statistical analysis of the Waldmeier and Gnevyshev-Ohl rules using the millennial-long sunspot-number series for the period 970 through 1900 including 85 solar cycles.

2. Data set

For the analysis we used the revised list of the reconstructed solar cycles from Usoskin et al. (2022) for the period 970 – 1900. Usoskin et al. (2021b) ascribed a quality flag Q to each reconstructed cycle: 0 – the cycle cannot be reliably identified; 1 – the cycle is greatly distorted; 2 – the cycle can be approximately identified, but its shape is distorted; 3 – a reasonably defined cycle; 4 – a well-defined cycle with a somewhat unclear amplitude; 5 – a clear cycle in both shape and amplitude. In the framework of this study, we applied the analysis primarily to the well-defined cycles (25 cycles with $Q \geq 4$) but discuss also reasonably defined (36 cycles with $Q \geq 3$) cycles. All analyses were applied to the annually-resolved data so that the ascending phase was defined with an accuracy of up to one year. The length of the ascending phase of a cycle was defined as the difference in full years between the year of the cycle minimum and the subsequent maximum as reported in Usoskin et al. (2022). Following the approach of Usoskin et al. (2021b), we considered, as the index of the solar-cycle strength, the cycle-averaged sunspot number $\langle SN \rangle$ which is more robustly defined than the maximum annual sunspot number. Direct sunspot-number data (ISN v.2) were not used at this stage to check whether the relations work for the reconstructed data and avoid bias.

3. Results

3.1. *Waldmeier relation*

The ascending phase $T_{\rm A}$ of solar cycles takes a length between 4 and 7 years. Accordingly, we made distributions of the cycle-averaged sunspot numbers $\langle SN \rangle$ for these values of $T_{\rm A}$ considering the uncertainties of the sunspot number reconstruction. This was done by using a Monte Carlo approach as described below. First, all well-defined ($Q \geq 4$) cycles with the given $T_{\rm A}$ were selected with their $\langle SN \rangle$ and $\sigma_{\langle SN \rangle}$. Next, using a pseudo-random number R normally distributed with zero mean and unit variance, we ascribed the exact value of the cycle-averaged SN with the uncertainties taken into account as

$$\langle SN \rangle^* = \langle SN \rangle + R \cdot \sigma_{\langle SN \rangle}. \tag{3.1}$$

The last step was repeated 1000 times and distributions of the $\langle SN \rangle^*$ values were collected in histograms, one for each $T_{\rm A}$ and fitted with a Gaussian as shown in Figure 2. As seen, the Gaussian fit describes the distribution reasonably well.

Figure 3 shows the scatter plot of the Waldmeier relation using the distributions of the $\langle SN \rangle$ shown in Figure 2. A clear tendency is observed that cycles with longer ascending phases are weaker. This scatter was fit by the linear relationship as shown by the solid line with 90% (two-side) confidence intervals.

The best-fit linear regression appears highly significant (significance $p < 0.01$) as

$$\langle SN \rangle = (-26 \pm 8) \cdot T_{\rm A} + (203 \pm 47), \tag{3.2}$$

where $T_{\rm A}$ is the length of the ascending phase in years. This is consistent with the regression reported by Usoskin et al. (2021b), viz. $\langle SN \rangle$=(-26±16)·$T_{\rm A}$+(197±90), but with smaller uncertainties, because of the corrected typo in the dataset. It is also fully consistent, within the uncertainties, with the Waldmeier rule based on the direct sunspot data (ISN v2.0, from SILSO) for the period since 1750 (Usoskin et al. 2021a), viz. $\langle SN \rangle$=(-20±9)·$T_{\rm A}$+(179±41) for the cycle-averaged sunspot numbers.

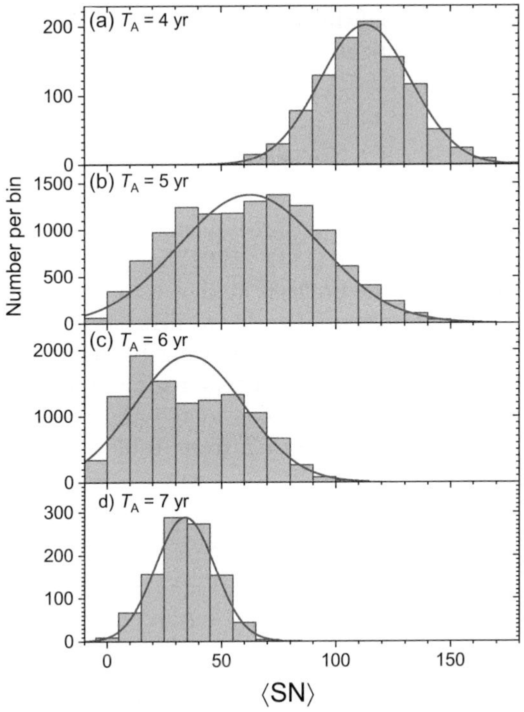

Figure 2. Distribution histograms of the $\langle SN \rangle$ values for the well-defined solar cycles ($Q \geq 4$) reconstructed from ^{14}C. Panels a through d correspond to the durations of the cycle ascending phase as $T_A = 4$, 5, 6 and 7 years, respectively. The red curves depict the best-fit Gaussians which are: a) 113 ± 20; b) 62 ± 31; c) 36 ± 24; d) 34 ± 13.

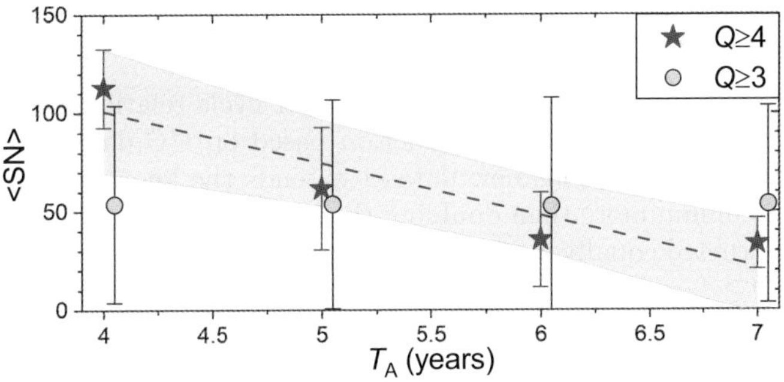

Figure 3. Relation between the cycle-average sunspot number $\langle SN \rangle$ and the length of the cycle ascending phase T_A for the millennium-long sunspot reconstruction. The red stars correspond to the distributions of well-defined cycles ($Q \geq 4$) shown in Figure 2, while the red dashed line with pink shading depicts the best-fit linear regression and its 90% (two-sided) confidence interval, respectively. The blue dots correspond to solar cycles with $Q \geq 3$.

We have also checked the validity of the Waldmeier rule using a larger statistic of poorly-defined cycles ($Q \geq 3$). The corresponding $\langle SN \rangle$ vs. T_A scatter is shown by blue dots in Figure 3. Even though the blue dots are consistent with the stars within uncertainties, there is no statistical relationship between the two variables, because of the too large error bars. Accordingly, the poorly-defined cycles spoil the empirical relation.

Thus, the Waldmeier rule is confirmed, using well-defined reconstructed cycles for the last millennium in the same quantitative shape as known for the last centuries from direct sunspot observations.

3.2. *Gnevyshev-Ohl rule*

We have also tried to check the validity of the Gnevushev-Ohl empirical rule using well-defined cycles ($Q \geq 4$), which form three clusters: (I) cycles 1 – 6 corresponding to years 976 – 1040 AD; (II) cycles 16 – 26 (1142 – 1262), and (III) cycles 56 – 59 (1584 – 1632). Since the absolute numbering of the cycles is unknown to define odd- and even-numbered ones, we have checked different options of cycle pairings within each cluster: the first cycle in the cluster is 'even' (pairing a) or 'odd' (pairing b). Thus, there are eight possible combinations of pairing cycles within the clusters, e.g., Ia + IIb + IIIa. We have checked all of them. As the cycle's intensity, we considered the product of the cycle-average sunspot number $\langle SN \rangle$ and the cycle length T from Table 1 of Usoskin et al. (2021b):

$$I = \langle SN \rangle \times T. \tag{3.3}$$

As the measure of the Gnevyshev-Ohl rule, we considered the difference in the intensities of the paired cycles, viz.

$$\Delta I_i = I_{i+1} - I_i. \tag{3.4}$$

The analysis was done also by applying a Monte-Carlo method as described in Section 3.1. We found that, for all possible combinations, ΔI appears indistinguishable from zero at the 1σ level ($p > 0.3$). This is likely because of the large uncertainties of the reconstructed sunspot numbers (of the order of a few hundred in units of ΔI) which are greater than the expected Gnevyshev-Ohl effect of about a hundred in units of ΔI as estimated from ISN (Usoskin et al. 2021a).

Thus, the Gnevyshev-Ohl rule cannot be confirmed using the new reconstructed solar cycle on the millennial timescale.

4. Conclusions

Here a quantitative analysis of the empirical solar-cycle relations has been performed using the updated solar-activity reconstruction based on ^{14}C data (Brehm et al. 2021; Usoskin et al. 2021b, 2022). This new dataset extends the known solar-cycle variability over the last millennium more than doubling the statistics of the cycles. However, not all cycles are reconstructed equally well. Here we mostly consider 25 well-defined cycles, with the quality flag $Q \geq 4$, and discuss ten more poorly-defined cycles with $Q = 3$. Non-defined cycles with $Q < 3$ are not considered.

We have shown that the Waldmeier rule saying that the cycle strength is inversely related to the length of the solar-cycle ascending phase, is fully confirmed on the millennial timescale using well-defined cycles. The use of poorly-defined cycles smears the relation. This implies that the Waldmeier rule is an intrinsic feature of the solar dynamo (e.g., Charbonneau 2020; Nandy 2021).

We have also checked the validity of the Gnevushev-Ohl rule, saying that solar cycles are tied in pairs of an even-numbered cycle followed by a slightly stronger odd-numbered cycle, on the millennial time scale. However, because of the relatively large uncertainties, the result is inconclusive, and the rule can be neither confirmed nor rebutted.

Acknowledgements

Partial support from the Academy of Finland (Projects ESPERA No. 321882) is acknowledged.

References

Aparicio, A. J. P., Vaquero, J., & Cruz Gallego, M. 2012, J. Space Weather Space Clim., 2, A12
Bard, E., Raisbeck, G., Yiou, F., & Jouzel, J. 1997, Earth Planet. Sci. Lett., 150, 453
Brehm, N., Bayliss, A., Christl, M., et al. 2021, Nature Geosci., 14, 10
Carrasco, V. M. S., Aparicio, A. J. P., Vaquero, J. M., & Gallego, M. C. 2016, Solar Phys., 291, 3045
Charbonneau, P. 2020, Living Rev. Solar Phys., 17, 4
Clette, F. & Lefèvre, L. 2016, Solar Phys., 291, 2629
Gnevyshev, M. N. & Ohl, A. 1948, Astron. Zh., 25, 18
Hathaway, D., Wilson, R., & Reichmann, E. 2002, Solar Phys., 211, 357
Hathaway, D. H. 2015, Living Rev. Solar Phys., 12, 4
Inceoglu, F., Simoniello, R., Knudsen, V. F., et al. 2015, Astron. Astrophys., 577, A20
Karoff, C., Inceoglu, F., Knudsen, M. F., Olsen, J., & Fogtmann-Schulz, A. 2015, Astron. Astrophys., 575, A77
Nandy, D. 2021, Solar Phys., 296, 54
Solanki, S., Krivova, N., Schüssler, M., & Fligge, M. 2002, Astron. Astrophys., 396, 1029
Takalo, J. & Mursula, K. 2018, Astron. Astrophys., 620, 100
Usoskin, I. G. 2017, Living Rev. Solar Phys., 14, 3
Usoskin, I. G., Kovaltsov, G., & Kiviaho, W. 2021a, Solar Phys., 296, 13
Usoskin, I. G., Mursula, K., Arlt, R., & Kovaltsov, G. A. 2009, Astrophys. J. Lett., 700, L154
Usoskin, I. G., Mursula, K., & Kovaltsov, G. A. 2001, Astron. Astrophys., 370, L31
Usoskin, I. G., Solanki, S. K., & Kovaltsov, G. A. 2007, Astron. Astrophys., 471, 301
Usoskin, I. G., Solanki, S. K., Krivova, N., et al. 2021b, Astron. Astrophys., 649, A141
Usoskin, I. G., Solanki, S. K., Krivova, N., et al. 2022, Astron. Astrophys., 664, C3
Waldmeier, M. 1935, Astron. Mitt. Z., 6, 105
Waldmeier, M. 1939, Astron. Mitt. Zurich, 14, 470

The Era of Multi-Messenger Solar Physics
Proceedings IAU Symposium No. 372, 2023
G. Cauzzi & A. Tritschler, eds.
doi:10.1017/S174392132300087X

Study of Ionospheric behaviour during the intense geomagnetic storms in 24th solar cycle over different Latitudes and Longitudes

Bhupendra Malvi[1], Sharad C. Tripathi[2] and P. K. Purohit[3]

[1]Department of Electronics, Barkatullah University, Bhopal - 462026, M.P., India

[2]School of Advanced Sciences and Languages (SASL), VIT Bhopal University, Bhopal - Indore Highway, Kothri Kalan, Sehore - 466114, M.P., India

[3]Department of Applied Sciences, NITTTR, Shamla Hills, Bhopal - 462003, M.P., India

Abstract. An analysis of geomagnetic disturbances and global ionospheric electron density perturbations during the 2015 St. Patrick's Day geomagnetic storm is presented in this paper. GPS observations from worldwide IGS stations are used and analysed through GPS-TEC analysis application developed by Gopi Seemala to get VTEC profiles. The St. Patrick's geomagnetic storm covers the interval of 15-23 March 2015, when transient solar eruptions (a prolonged C9-class solar flare and associated CMEs on 15 March) and a strong geomagnetic storm during 16-18 March (Dst dropped to -223 nT) were reported. This geomagnetic storm led to complex effects on the ionosphere. The global maps have been created after analysing VTEC profiles at Low, Mid and High-latitudes over different longitudinal sectors. Major features of the positive and negative ionospheric storm development are observed in Asian, European and American Low, Mid and High-latitudes.

Keywords. Geomagnetic Storm, Total Electron Content, Ionospheric Perturbations, St. Patrick's Day

1. Introduction

The biggest geomagnetic storm of the 24th solar cycle occured in the middle of March 2015 with the value of Dst index at -223 nT. It was named as the "St Patrick's Day" storm due to the occurence of its main phase on 17 March. This storm was associated with a Coronal Mass Ejection occured on 15 March 2015. SOHO/LASCO C3 recorded this partial halo CME and an associated C class solar flare was observed which erupted from the active region located at S22W25. A series of type II/IV radio bursts were also observed. It is well known that the ionospheric electric field disturbances that are observed in the medium and low latitude regions during a magnetic storm have different timescales and originate from the direct penetration of the magnetospheric electric field into the ionosphere (Nishida 1968; Vasyliunas 1970; Jaggi & Wolf 1973; Fejer et al. 1990). Magnetospheric, ionospheric and atmospheric changes during geomagnetic storms are fundamentally dynamic, interactive and nonlinear which can be represented by variations in key parameters, such as ionospheric electron densities, total electron content, electromagnetic field and currents, particle precipitation etc. The essential physical processes that can explain these coupled interactions are known only up to a certain degree, with several qualitative explanations. However, the relative importance varies from case to case establishing different storm time responses in the geospace (Zhang et al. 2017).

The present study is the first of its kind, where the ionospheric perturbations during the St. Patrick's Day storm of 17 March 2015 (which has a long-lasting main phase with

Table 1. List of the used IGS Stations.

IGS Stations					
Southern Hemisphere			Northern Hemisphere		
Lat	**Long**	**Station Code**	**Lat**	**Long**	**Station Code**
−77.83	166.66	MCM4	0.35	9.67	NKLG
−69	39.58	SYOG	5.25	307.19	KOUR
−68.57	77.97	DAV1	12.14	273.75	MANA
−66.28	110.51	CAS1	13.02	77.57	IISC
−64.77	295.94	PALM	13.43	144.8	GUUG
−63.32	302.09	OHI3	13.73	100.53	CUSV
−53.13	289.12	PARC	20.7	203.74	MAUI
−49.35	70.25	KIT3	21.85	257.71	INEG
−43.73	183.38	CHTI	30.59	34.76	RAMO
−37.82	144.97	MOBS	32.64	343.09	FUNC
−34.9	302.06	IPGS	35.14	33.39	NICO
−32.38	20.81	SUTH	35.95	140.65	KSMV
−31.86	133.8	CEDU	36.13	138.36	USUD
−29.04	115.34	YAR2	37.74	334.33	PDEL
−27.125	250.65	ISPA	39.02	283.17	GODZ
−25.89	27.68	HRAO	42.99	74.75	CHUM
−22.57	17.08	WIND	48.13	240.31	BREW
−20.27	289.86	IQQE	59.57	150.77	MAG0
−19.26	147.05	TOW2	60.53	29.78	STVL
−19.07	190.07	NIUM	60.75	224.77	WHIT
−12.93	321.56	SAVO	64.13	338.04	REYK
−12.18	96.83	COCO	64.26	344.81	HOFN
			64.97	212.5	FAIR
			69.36	88.35	NRIL
			71.63	128.86	TIXI
			74.69	265.1	RESO

the Dst as low as -223 nT), have been analysed all over the globe, altogether. Strong equatorial zonal electric field enhancement were observed due to two strong prompt penetration electric field phases during the main phase of this storm. The first one occured around 1200 UT with the southward IMF Bz and the second one around 1725 UT with the onset of a substorm. The enhanced electric field due to this prompt penetration of electric field lead to the strong super fountain effect (Venkatesh et al. 2017). The consequent spatiotemporal disturbances in the ionospheric TEC profile becomes more important to be investigated.

2. Datasets and Methodology

In the present analysis GPS data from various IGS stations, located around the globe, have been downloaded from Scripps Orbit and Permanent Array Center (SOPAC) website `http://sopac.ucsd.edu/` and analysed using GPS-TEC analysis software developed by Gopi Seemala [`http://seemala.blogspot.com/2017/09/gps-tec-program-ver-295.html` to get VTEC (Vertical Total Electron Content), which is the vertically integrated electron density at a particular location at a particular time, profiles. Dst index data were downloaded from World Data Center (`http://wdc.kugi.kyushu-u.ac.jp`). Geomagnetically quiet days were identified from World Data Center repository and the respective GPS data were downloaded for the said IGS stations. One day VTEC profile was generated from these quiet day observations by taking standard deviations of respective observational instants on these days. This prepared data (3-sigma) was used as background and subtracted from the VTEC profiles of disturbed days; during 15-23 March 2015. Peak values on the said days, for all the IGS stations, are used to create global TEC maps. The list of IGS stations considered in present analysis are listed in table 1.

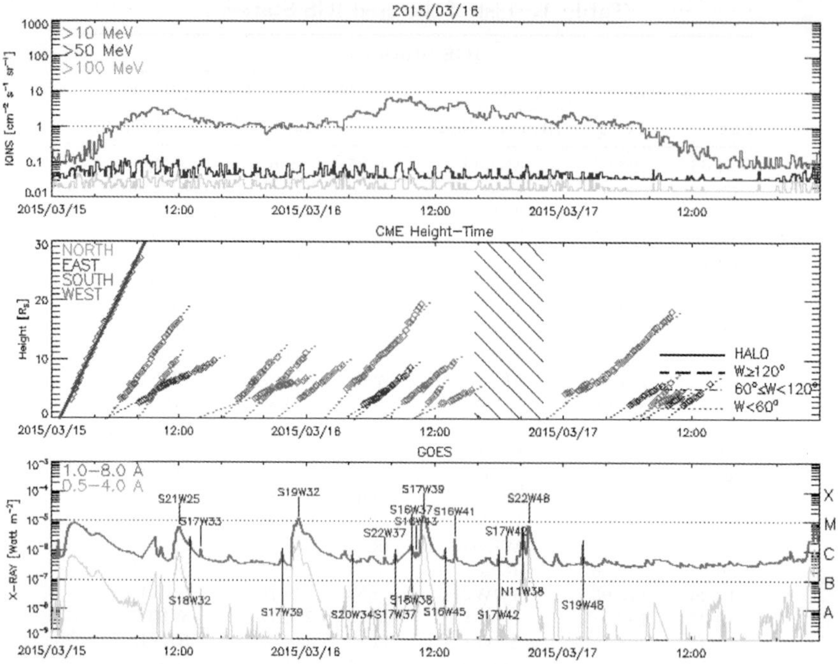

Figure 1. Solar Transients associated to St. Patrick's Day Geomagnetic Storm.

3. Results and Discussion

An intense geomagnetic storm was observed on March 17, 2015 having Dst as strong as -223 nT at 10:00 hours UT. A total of 48 stations were identified, from all over the globe, for which GPS data was availabile during the observational window 15-23 March 2015. These stations were selected for analysing ionospheric perturbations in various longitudinal and latitudinal regions; represented as grids in the figure 3. Several solar transients were observed before the main St. Patrick's Day event, and are possibly associated with it, see figure 1 [Source: `https://cdaw.gsfc.nasa.gov/`]. Dst data downloaded from World Data Center repository is plotted in figure 2. Global VTEC profiles are shown in figure 3 where daily background corrected peak values are plotted in the various grids of subplots (from a to i). In the 0°-30°E longitude range it is observed that the TEC is increased in the equatorial and mid latitude regions of southern hemisphere during the main phase of geomagnetic storm. In the 30°E-60°E longitude range peak values of background corrected daily VTEC profiles are plotted in the global TEC maps and it is observed that the TEC was increased everywhere, except at the high latitude station in southern hemisphere, during the main phase of the geomagnetic storm. In the 60°E-90°E longitude range it is observed that the TEC decreased everywhere during the main phase of the geomagnetic storm. In the 90°E-120°E longitude range it is observed that the TEC decreased everywhere during the main phase of the geomagnetic storm. In the 120°E-150°E longitude range it is observed that the TEC decreased everywhere during the main phase of the geomagnetic storm. In the 150°E-180° longitude range it is observed that the TEC increased everywhere during the main phase of the geomagnetic storm. In the 150°W-180° longitude range it is observed that the TEC decreased, except at the equatorial station in the southern hemisphere, during the main phase of the geomagnetic storm. In the 120°W-150°W longitude range it is observed that the TEC increased, except at the higher latitude station in northern hemisphere, during the main phase of the geomagnetic storm and decreased during the recovery phase. Equatorial region stations in both the

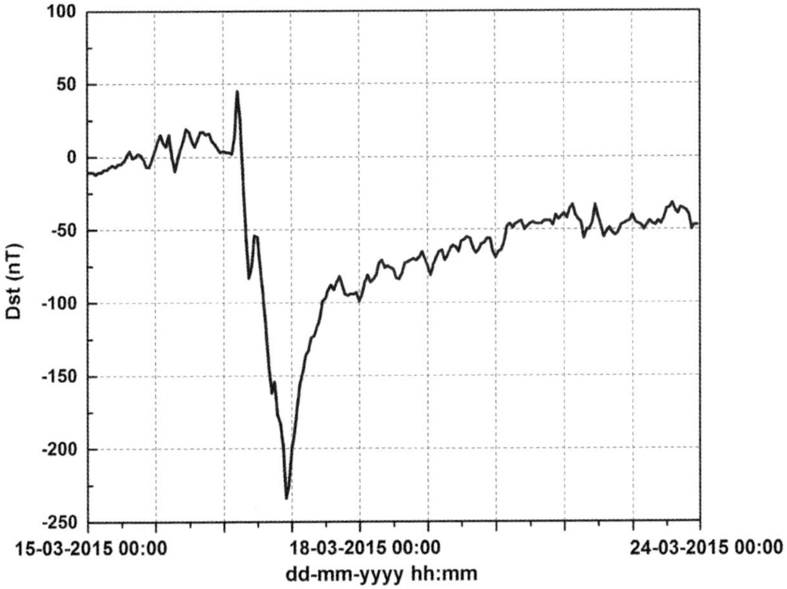

Figure 2. Dst variations during St. Patrick Day Geomagnetic Storm.

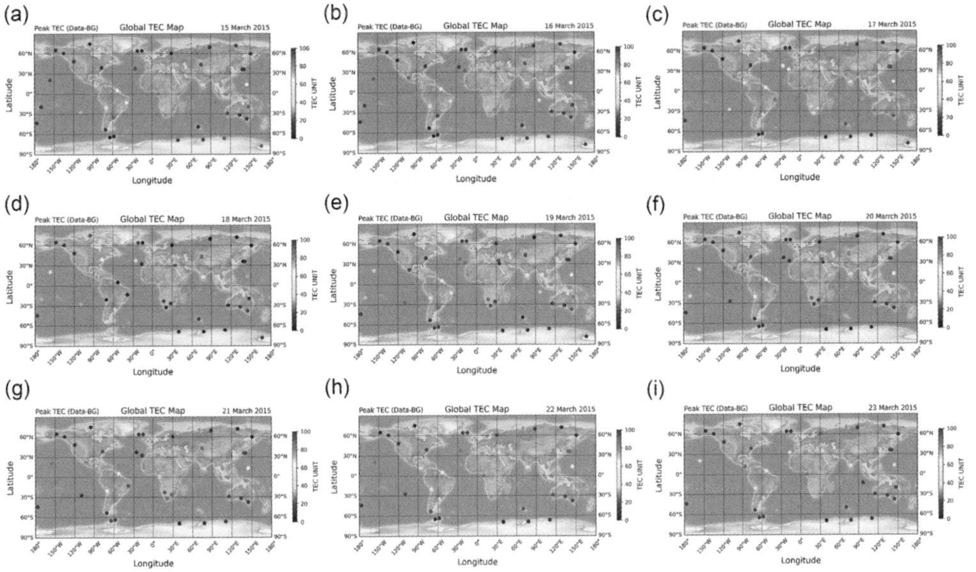

Figure 3. Global Ionospheric responses during St. Patrick's Day Geomagnetic Storm.

hemispheres and Mid latitude station in the southern hemisphere witnessed most prominent disturbance during the main phase of geomagnetic storm, whereas, the mid latitude station in the northern hemisphere got delayed response, ionospheric TEC increased here, next day. In the 90°W-120°W longitude range it is observed that the TEC increased, except for the higher latitude station in northern hemisphere, during the main phase of the geomagnetic storm and decreased during the recovery phase. Mid latitude station in both the northern and southern hemisphere witnessed most prominent disturbance during the main phase of geomagnetic storm. In the 60°W-90°W longitude range it is observed that the TEC increased during the main phase of the geomagnetic storm and decreased during the recovery phase. Mid latitude station in the northern hemisphere witnessed most prominent disturbance during the main phase of geomagnetic storm. In the 30°W-60°W longitude range it is observed that the TEC increased during the main phase of the geomagnetic storm, in the equatorial region and higher latitudes of southern hemisphere. In the 0°-30°W longitude range it is observed that the TEC increased during the main phase of the geomagnetic storm, in the mid latitude and higher latitudes regions of northern hemisphere.

4. Conclusions

Global ionospheric perturbations in VTEC profiles at various longitudinal regions reflect delayed effects at some locations, somewhere it is positive and somewhere it is negative, due to the complex geomagnetic responses. The interaction of the Solar Plasma at geospace and the storm preparation is believed to plays a key role in such complex disturbances.

Acknowledgements

Author Bhupendra Malvi is thankful to the International Astronomical Union (IAU) for providing a grant for his participation in IAUGA 2022 hosted at BEXCO in Busan, Republic of Korea during August 2-11, 2022.

References

Fejer, B., Spiro, R., Wolf, R., & Foster, J. 1990, Geophysical Journal;(France), 8

Jaggi, R. K. & Wolf, R. A. 1973, Journal of Geophysical Research (1896-1977), 78, 2852

Nishida, A. 1968, Journal of Geophysical Research (1896-1977), 73, 5549

Vasyliunas, V. M. 1970, in Particles and Fields in the Magnetosphere, ed. B. M. McCormac (Dordrecht: Springer Netherlands), 60–71

Venkatesh, K., Tulasi Ram, S., Fagundes, P. R., Seemala, G. K., & Batista, I. S. 2017, Journal of Geophysical Research: Space Physics, 122, 4553

Zhang, S.-R., Zhang, Y., Wang, W., & Verkhoglyadova, O. P. 2017, Journal of Geophysical Research (Space Physics), 122, 6901

The Era of Multi-Messenger Solar Physics
Proceedings IAU Symposium No. 372, 2023
G. Cauzzi & A. Tritschler, eds.
doi:10.1017/S1743921323000510

Power spectra of the Sun's large-scale magnetic field during solar cycles 23 and 24

Yukun Luo[1,2], Jie Jiang[1,2] and Ruihui Wang[1,2]

[1]School of Space and Environment, Beihang University, Beijing, China

[2]Key Laboratory of Space Environment monitoring and Information Processing of MIIT, Beijing, China

Abstract. The magnetic power spectrum analysis provides an effective way to understand the observed distribution of the photospheric magnetic fields and their interaction with plasma motions. We aim to investigate the power spectra using spherical harmonic decomposition of SOHO/MDI and SDO/HMI synoptic magnetograms for cycles 23 and 24. We find that the calibration factor between MDI and HMI power spectral density is spatial scale-dependent. The magnetic power spectra show two peaks at the AR scale (l≈30) and supergranular scale (l≈120). The power law indices between these show a good anti-correlation with the amplitude of magnetic activity.

Keywords. Sun: magnetic fields, Sun: activity, Sun: photosphere

1. Introduction

People usually utilize Fourier decomposition of local magnetograms to derive the magnetic power spectrum and further investigate the properties of the photospheric magnetic fields, especially the small-scale field. However, the long-term evolution of the large-scale magnetic field is rarely analyzed based on the power spectrum. Also, there are questions of what is the proper calibration factor to adopt when doing comparative studies of HMI and MDI, e.g., Liu et al. (2012). So this work aims to get new calibration method and analyze the power spectrum across two solar cycles.

2. Methods

Prior to the analysis, we pre-processed the synoptic magnetograms. Then the spherical harmonic decomposition were conducted on these data to obtain the spherical harmonic coefficients, which are used to calculate the power spectra based on Parseval's theorem.

3. Results

We found that the calibration function between MDI and HMI power spectral density (B^2_{MDI}/B^2_{HMI}) is $-0.0021 * l + 1.91$ rather than 1.42^2 (Fig. 1). After the spatial scale-dependent calibration, MDI and HMI show roughly the same power spectrum distributions in the range from $l = 5$ to $l = 200$ and a distinct peak at about $l \approx 120$.

Significant power is found at two spatial scales, i.e., about $l = 30$ (146 Mm AR size) and $l = 120$ (36 Mm supergranular size), when analyzing the total magnetic field and non-axisymmetric component. No such features are instead observed in the spectra of the axysimmetric component of the field, which shows a continuous decrease (Fig. 2). The power-law indices (k) for the spatial range of $l = 30$ and $l = 120$ show a good anti-correlation with the amplitude of magnetic activity. The function between them is

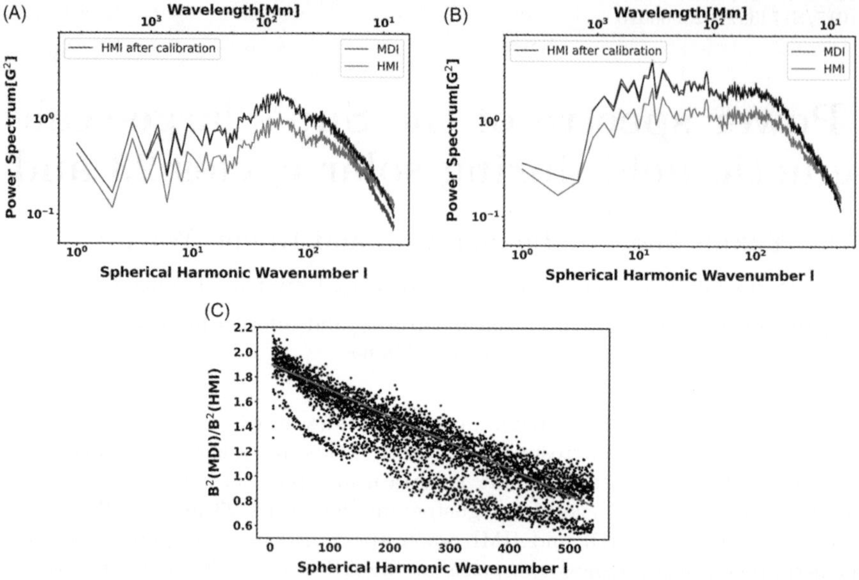

Figure 1. Comparison and calibration of the magnetic power spectra obtained by MDI and HMI maps. CRs 2097 (panel A) and 2100 (panel B) are taken as examples to illustrate the power spectra of the HMI map (red), MDI map (blue), and HMI map calibrated (black) using the red line in Panel (C). Panel (C): spatial scale dependence of the calibration factor between MDI and HMI power spectral density. Black dots and red lines are the observations and the linear fit to the data, respectively.

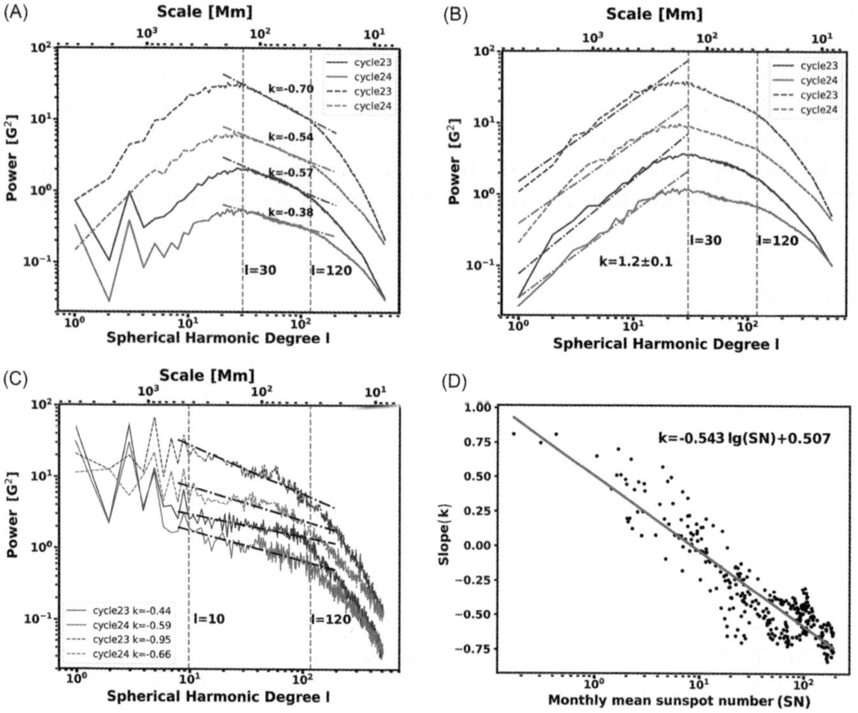

Figure 2. Magnetic power spectra for total magnetic field (panel A), for non-axisymmetric field (m≠0, panel B), and for axisymmetric field (m=0, panel C) of cycles 23 and 24. Red (blue) curves: cycle 24 (23); dotted (solid) curves: active (quieter) phase of a cycle. Panel D: Power-law indices k shown in panel A of the power spectra as a function of the sunspot number.

$k = -0.53lg(SN) + 0.51$. For the non-axisymmetric field, the spectrum for the scales larger than ARs' size shows a constant power index, that is 1.2.

4. Conclusions

The magnetic power spectra over broad spatial scales from about 22 Mm to the global scale for cycles 23 and 24 are derived and analyzed. We obtained a new calibration function between MDI and HMI maps. The power spectra reflect the characteristics related to solar activity and solar structures such as supergranlues.

Acknowledgements

The SDO/HMI data are courtesy of NASA and the SDO/HMI team. SOHO is a project of international cooperation between ESA and NASA. The research is supported through the National Natural Science Foundation of China (grant Nos. 12173005 and 11873023).

Supplementary material

To view supplementary material for this article, please visit https://doi.org/10.1017/S1743921323000510.

Reference

Liu, Y., Hoeksema, J. T., Scherrer, P. H., et al. 2012, Sol. Phys., 279, 295

The Era of Multi-Messenger Solar Physics
Proceedings IAU Symposium No. 372, 2023
G. Cauzzi & A. Tritschler, eds.
doi:10.1017/S1743921323000467

Long-term Evolution of Magnetic Field Proxies as Deduced from Archival Data

Dipankar Banerjee[1,2], Bibhuti Kumar Jha[1,2], Sudip Mandal[2], Subhamoy Chatterjee[3], Manjunath Hegde[1] and KoSO Digitization Team

[1]Arybhatta Research Institute of Observational Sciences, Nainital-263001, India
Email: *dipu@aries.res.in*

[2]Indian Institute of Astrophysics, Bangalore-560034, Karnataka

[3]Max Planck Institute for Solar System Research, Göttingen, Germany

[4]Southwest Research Institute, 1050 Walnut Street #300, Boulder, CO, 80302, USA

Abstract.
Regular observations of the solar magnetic field are available only for about the last five cycles. Thus, to understand the origin of the variation of the solar magnetic field, it is essential to reconstruct the magnetic field for the past cycles, utilizing the proxies of the magnetic field from other data sets. Long-term uniform observations for the past 100 yrs, as recorded at the Kodaikanal Solar Observatory (KoSO), in multi-wavelengths provide such an opportunity. Various automatic techniques have been developed to extract these features from KoSO data. We analyzed the properties of these extracted features to understand global solar magnetism in the past.

Keywords. Sun: magnetic fields, Sun: photosphere, Sun: chromosphere, catalogs

1. Introduction

The unavailability of solar magnetic field measurement before the 1970s makes the historical records of solar observation a crucial resource to understand the physics behind the solar activity and its long-term variability. Various solar features such as sunspots, plages, filaments/prominence etc., as observed in the multi-wavelength observation of the Sun, serve as proxies of the solar magnetic field. Therefore, understanding these features over the century will help us understand global solar magnetism over such a long period. In this sense, Kodaikanal Solar Observatory (KoSO) in India has a record of historical observation of the Sun, simultaneously in white light (1904 – 2017), Ca K (1904 – 2007) and H-alpha (1912 – 2007), over the century. In Figure 1, we show the number of observations per year for all these data sets. These data were initially taken on the photographic plates/films and have been digitized now, using 4k × 4k CCD, and made available for the community.

2. A Few Results from KoSO Digital Archive

The state-of-the-art techniques has been developed to detect the sunspot (Ravindra et al. 2013; Mandal et al. 2017; Jha et al. 2022), sunspot umbra (Jha et al. 2019), plages (Chatterjee et al. 2016), filaments (Chatterjee et al. 2017) and Ca-K prominence (Chatterjee et al. 2020) from these data sets. The latitude-time plot for all these data sets are shown in Figure 2a (sunspot), 2b (plages), 2c (filaments) and 2d (Ca-k

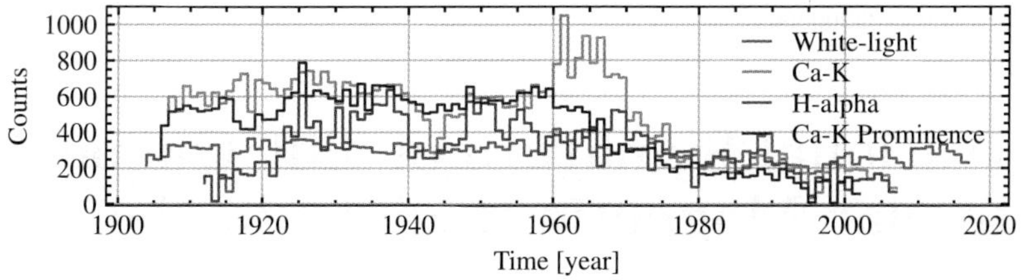

Figure 1. Number of observation per year as available in the KoSO digital archive.

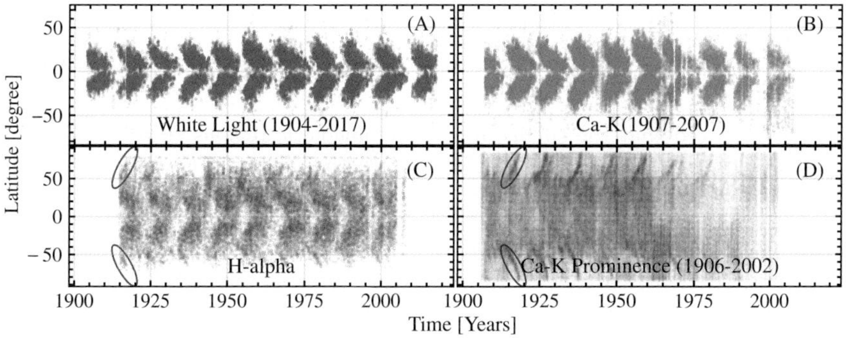

Figure 2. The butterfly diagram generated from (A) white-light, (B) Ca-k, (C) H-alpha and (D) Ca-k prominence. In Panel-(C) and (D) red ellipses are representing the pole-ward migration of filaments/prominence.

prominences). These digitized data sets have been extensively used (Mandal et al. 2017, 2020; Jha et al. 2019, 2021) to study the various aspects of long-term solar variability. Recently, Mordvinov et al. (2020) have reconstructed the magnetic butterfly diagram corresponding to the past cycle using the digitized data from KoSO.

Acknowledgements

DB and BKJ would like to thank IAU for providing the travel support to attend the IAUGA 2022 in-perosn in South Korea. All these data are accessible through `https://kso.iiap.res.in/new`.

Supplementary material

To view supplementary material for this article, please visit https://doi.org/10.1017/S1743921323000467.

References

Chatterjee, S., Banerjee, D., & Ravindra, B. 2016, ApJ, 827, 87

Chatterjee, S., Hegde, M., Banerjee, D., & Ravindra, B. 2017, ApJ, 849, 44

Chatterjee, S., Hegde, M., Banerjee, D., Ravindra, B., & McIntosh, S. W. 2020, Earth and Space Science, 7, e00666

Jha, B. K., Hegde, M., Priyadarshi, A., et al. 2022, Frontiers in Astronomy and Space Sciences, 9, 1019751

Jha, B. K., Mandal, S., & Banerjee, D. 2019, *Solar Phys.*, 294, 72

Jha, B. K., Priyadarshi, A., Mandal, S., Chatterjee, S., & Banerjee, D. 2021, *Solar Phys.*, 296, 25

Mandal, S., Hegde, M., Samanta, T., et al. 2017, A&A, 601, A106

Mandal, S., Krivova, N. A., Solanki, S. K., Sinha, N., & Banerjee, D. 2020, A&A, 640, A78

Mordvinov, A. V., Karak, B. B., Banerjee, D., et al. 2020, ApJ Letters, 902, L15

Ravindra, B., Priya, T. G., Amareswari, K., et al. 2013, A&A, 550, A19

The Era of Multi-Messenger Solar Physics
Proceedings IAU Symposium No. 372, 2023
G. Cauzzi & A. Tritschler, eds.
doi:10.1017/S1743921322005002

Observational Signature of Tilt Quenching of Bipolar Magnetic Regions in the Sun

Bibhuti Kumar Jha[1], **Bidya Binay Karak**[2] **and Dipankar Banerjee**[1]

[1]Arybhatta Research Institute of Observational Sciences, Nainital-263001, India
Email: *bibhuti@aries.res.in*

[2]Department of Physics, Indian Institute of Technology (BHU), Varanasi 221005, India

Abstract. It is believed that the tilt in the bipolar magnetic regions (BMRs) is produced due to a torque induced by the Coriolis force, acting on the diverging flow from the apex of the rising flux tube of the toroidal field in the solar convection zone (SCZ).The BMRs with a strong magnetic field are expected to have reduced tilt as they rise very quickly in the SCZ. This effect can provide the required nonlinear quenching mechanism to suppress the growth of magnetic filed in the dynamo models. Here, we use the magnetograms of the Michelson Doppler Imager (1996–2011) and Helioseismic and Magnetic Imager (2010–2018) to automatically detect the BMRs and look for the signature of tilt quenching. Based on the Bayesian inference method, our results show that the posterior distribution of quenching parameters is Gaussian, and the mean of this distribution agrees with the earlier findings.

Keywords. Sun: magnetic fields, (Sun:) sunspots, Sun: photosphere, methods: statistical

1. Introduction

In the thin flux tube model, a rising flux tube experiences Coriolis force in the SCZ, appearing as a tilted Bipolar Magnetic Region (BMR) on the photosphere. Hale et al. (1919) observed that the tilt (γ) of BMRs depends on the latitude (λ) and given as $\gamma = \gamma_0 \sin\lambda$, which is know as the Joy's law. These tilted BMRs play a vital role in the Babcock–Leighton process (Babcock 1961; Leighton 1964) for the generation of the poloidal field (Mordvinov et al. 2022). In the kinematic dynamo models such as the Babcock-Leighton type, the saturation of the magnetic field is a big concern, and most of the modelers take an ad-hock nonlinear quenching of type

$$\gamma = \frac{\gamma_0 \sin\lambda}{1 + \left(\frac{B}{B_0}\right)^n} \tag{1.1}$$

to suppress the growth of magnetic field (Lemerle & Charbonneau 2017; Karak & Miesch 2017, 2018, with $n = 2$). Here B_0 is the magnetic field limit, beyond which we can expect the effect of tilt quenching. In contrast with the findings of Stenflo & Kosovichev (2012), Jha et al. (2020) reported an indication of tilt quenching and estimated the value of B_0 and n. Here in this proceedings, we have extended the work of Jha et al. (2020) and used Bayesian inference to better estimate these parameters.

2. Data and Method

We use the MDI (1996 – 2011) and HMI (2010 – 2018) magnetogram data and detected the BMRs using the same automatic technique as Jha et al. (2020) and

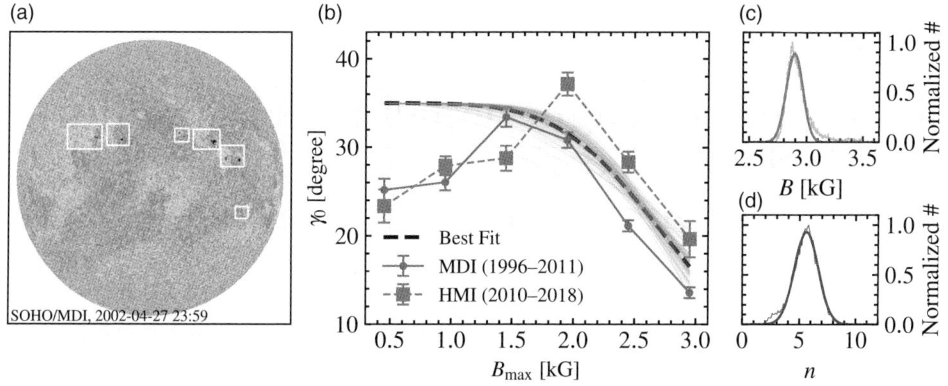

Figure 1. (a) An example showing the identified BMRs from MDI data. (B) Variation of γ_0 as a function of $B_{\max}$, light grey and dashed black lines shows Equation 1.1 based on the 500 randomly selected parameters from the distribution and the mean of the distribution. (c) and (d) show the posterior distribution of $B_{\max}$ and n.

Stenflo & Kosovichev (2012). A representative example of BMR detection is shown in Figure 1a.

3. Results and Conclusion

In Figure 1, we plot the γ_0 as a function of $B_{\max}$ for both data set (MDI and HMI), where we can see a signature of tilt quenching in high field regime ($B_{\max}$> 2 KG) similar to findings of Jha et al. (2020). We use the observed data as prior information to estimate model parameters (B_0 and n), called posterior distribution, using the Bayesian inference method. Assuming Gaussian likelihood and uniform prior distribution, we calculated the posterior distribution for these two parameters (B_0 and n), which is shown in Figure 1c and 1d. In Figure 1b, we plot Equation 1.1 using randomly selected 500 parameter sets from the distribution (light grey lines) and the mean of the distribution based on Gaussian fitting (dashed black line). The mean values, $\overline{B}_0 = 2893 \pm 73$ G and $\overline{n} = 5.6 \pm 1.1$, agree with the findings of Jha et al. (2020) based on least-square fit and indicate an even more substantial quenching effect. Readers may go through this YouTube link to watch the details about this work.

Acknowledgements

Authors would like to thank IAU for providing the travel support to attend the IAUGA 2022 in-person in South Korea. BKJ woukd also like to thank ARIES Nainital for taking care of rest of the expences.

References

Babcock, H. W. 1961, ApJ, 133, 572
Hale, G. E., Ellerman, F., Nicholson, S. B., & Joy, A. H. 1919, ApJ, 49, 153
Jha, B. K., Karak, B. B., Mandal, S., & Banerjee, D. 2020, ApJ Letters, 889, L19
Karak, B. B. & Miesch, M. 2017, ApJ, 847, 69
Karak, B. B. & Miesch, M. 2018, ApJ Letters, 860, L26
Leighton, R. B. 1964, ApJ, 140, 1547
Lemerle, A. & Charbonneau, P. 2017, ApJ, 834, 133
Mordvinov, A. V., Karak, B. B., Banerjee, D., et al. 2022, MNRAS, 510, 1331
Stenflo, J. O. & Kosovichev, A. G. 2012, ApJ, 745, 129

The Era of Multi-Messenger Solar Physics
Proceedings IAU Symposium No. 372, 2023
G. Cauzzi & A. Tritschler, eds.
doi:10.1017/S1743921322005026

Photospheric magnetic field variation during solar flares and their implication for the generation of sunquakes

S. Vargas Domínguez[1], W. Caicedo Tez[1] and J. C. Buitrago Casas[2]

[1]Universidad Nacional de Colombia, Observatorio Astronómico Nacional

[2]Space Sciences Laboratory, University of California, Berkeley

Abstract. Solar flares are an explosive manifestation of the complex magnetic structuring of active regions in the solar atmosphere. The photospheric magnetic field is found to change rapidly, abruptly, and significantly during flaring events. Previous studies are mainly based on line-of-sight or low-cadence data. In this work, we focus on the temporal and spatial evolution of the permanent changes in the magnetic field of solar flares from high-cadence vector data (135 seconds) of the imaging system (dopplergrams and magnetograms) of the SDO/HMI instrument. The highly energetic events under analysis occurred during the solar cycle 24, covering low and high energy ranges, according to GOES classification. This investigation also stands as a crucial input for the characterization and understanding of sunquakes.

Keywords. Sun: activity, Sun: flares, Sun: photosphere

1. Introduction

During a flaring event, the magnetic field topology changes rapidly, abruptly, and significantly. Some of these eruptive events inject enough energy into the photosphere and sub-photosphere to generate acoustic responses observed as sunquakes. The precise physical mechanism causing the acoustic source of a sunquake is still a topic of debate. Most authors agree that magnetic field restructuring must play a fundamental role in causing such acoustic drivers. Previous studies have mainly probed the line-of-sight component of the magnetic field in such scenarios. In this work, we investigate the temporal and spatial evolution of permanent changes in the magnetic field geometry.

2. Data

We use a sample of six acoustically active flaring events selected from Buitrago-Casas et al. (2015); Sharykin and Kosovichev (2020), as listed in Table 1. We used vector magnetograms acquired with the Solar Dynamics Observatory (SDO) with the Helioseismic and Magnetic Imager (HMI) instrument. The highly energetic events under study occurred during the past solar cycle 24, and cover a range of high and low GOES classes.

3. Analysis and Results

In order to probe the geometry of the magnetic field, we applied a procedure starting with choosing an appropriate interval covering the impulsive phase of each flaring event and identifying the region of interest. Maps of magnetic inclination differences are then constructed. Figure 1 shows an example of a region with the map of inclination with

Table 1. List of the studied flaring events.

Number	Flare	Heliospheric position	AR
1	SOL20130217T15:50-M1.2	[-338,307]	11675
2	SOL20131106T13:49-M3.8	[-549,267]	11890
3	SOL20131107T03:40-M2.3	[-450,-272]	11890
4	SOL20120509T14:02-M1.8	[-351,159]	11476
5	SOL20170904T15:11-M1.5	[-490,252]	12673
6	SOL20110213T17:28-M6.6	[-37,-132]	12297

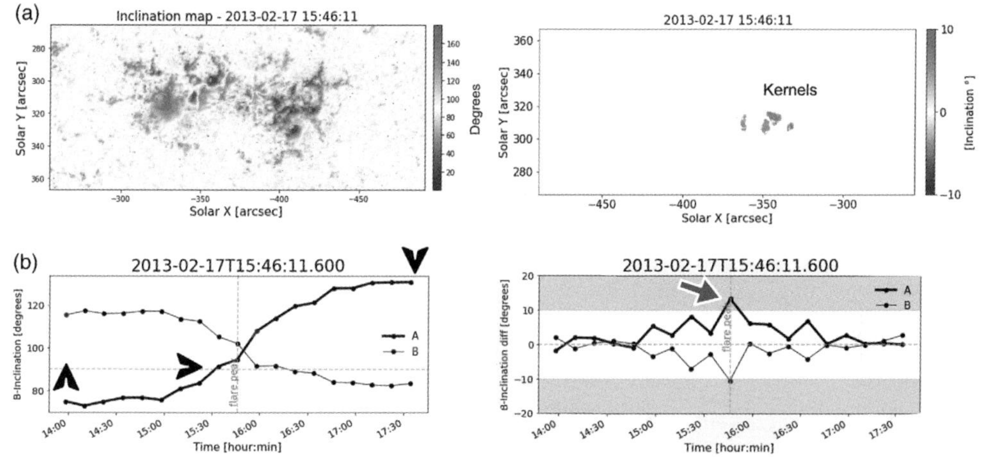

Figure 1. Example of maps and plots constructed to study the evolution of the magnetic field inclination for active region number 1 in Table 1. See text for details. This research used the SunPy 4.0.5 package (Mumford et al. 2020).

identified kernels (top panels) and plots for the time evolution of magnetic field inclination and inclination differences (bottom panels). Changes of tens of degrees in the magnetic field inclination are present in the flaring events displaying sunquakes. The acoustic wave is produced in the lower atmosphere by the ponderomotive force in the Alfvénic front associated with the magnetic variation.

4. Take-away ideas

Below we summarize some of the main ideas extracted from this work:

- Permanent changes in the magnetic field inclination with values greater than 10° in a kernel are likely to induce sunquake signatures, as stated in Russell et al. (2016).
- Results are in agreement with theoretical approaches.
- Injection of particles can not take place if the field lines do not display the correct direction (downwards).
- Sunquakes offer an opportunity to study the interaction of acoustic waves with magnetic fields and flows in flaring regions.

Supplementary material

To view supplementary material for this article, please visit https://doi.org/10.1017/S1743921322005026.

References

Buitrago-Casas, J. C., et al., 2015, Solar Physics 290.11: 3151-3162.
Mumford, S., Freij, N., Christe, S., et al. 2020, The Journal of Open Source Software, 5, 1832.
Russell, A. J. B., Mooney, M. K., Leake, J. E., & Hudson, H. S. 2016, ApJ, 831, 42.
Sharykin, Ivan N., and Alexander G. Kosovichev., 2020, ApJ, 895.1: 76.

The Era of Multi-Messenger Solar Physics
Proceedings IAU Symposium No. 372, 2023
G. Cauzzi & A. Tritschler, eds.
doi:10.1017/S1743921323000455

A new mechanism for the butterfly diagram of the solar cycle

Zebin Zhang[1,2], Jie Jiang[1,2] and Haowei Zhang[1,2]

[1]School of Space and Environment, Beihang University, Beijing, China

[2]Key Laboratory of Space Environment monitoring and Information Processing of MIIT, Beijing, China

Abstract. The butterfly diagram of the solar cycle is the equatorward migration of sunspot's emergence latitudes as the solar cycle evolves, which was attributed to the equatorward flow at the base of the convection zone. However, helioseismological studies indicate controversial forms of the flow, and even present poleward flow at the base, which poses a big challenge to the wide-accepted mechanism. So we aim to propose a new mechanism, that is the latitude-dependent radial flux transport.

Keywords. Sun: activity, Sun: magnetic fields, sunspots

1. Introduction

Revealing the mechanism causing the butterfly diagram is essential for the understanding of the solar and stellar dynamo. The mechanism that meridional flow causes the butterfly diagram is well-accepted but faces challenges from helioseismological results (Zhao et al. 2013). So new mechanisms are needed. It was suggested that dynamo wave in the near-surface shear layer and latitudinal pumping are two potential candidates (Guerrero & de Gouveia Dal Pino 2008; Pipin & Kosovichev 2011). However, these candidates all have their limits. So this work aims to propose a new mechanism from a new perspective.

2. New mechanism

2.1. *Model*

We adopt the data-driven FTD model of Jiang et al. (2013) to carry out this work and keep most of the ingredients unchanged except the meridional flow and radial pumping. In this work we consider three typical profiles adopted by literature, whose latitudinal velocities are shown in Figure 1. And we consider two profiles of the radial pumping. One is latitude-independent, which has the same strength at all latitudes. The another one is latitude-dependent, whose strength becomes stronger as the latitudes increase.

2.2. *Simulation results and illustration of the new mechanism*

In this section, we will show that the latitude-dependent radial transport of the poloidal flux causes the equatorward latitudinal migration of the toroidal fields at the base of the convection zone (CZ), under different profiles of meridional flows. The results are shown in Figure 2.

Figure 2(a) shows a poleward migration of the toroidal fields at low latitudes. This migration is only caused by the poleward dynamo wave since there is no flow at the

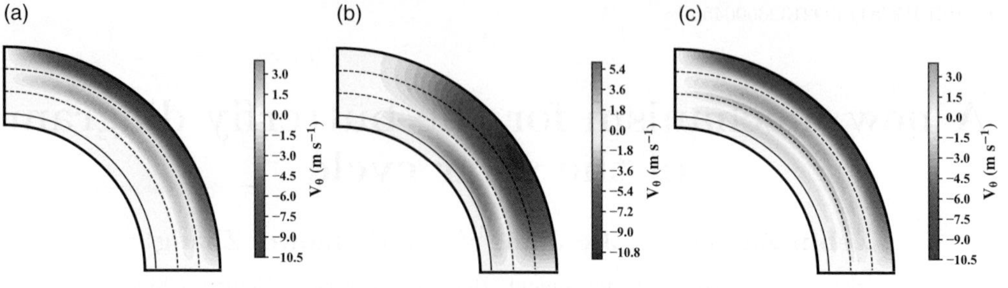

Figure 1. latitudinal velocities of meridional flow profiles used in this work. (a) Shallow single cell. (b) Deep single cell. (c) Double cell.

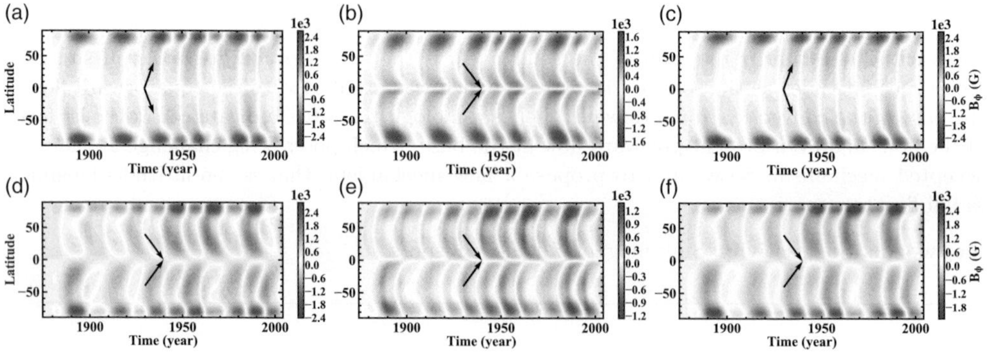

Figure 2. Time-latitude diagrams of the toroidal field at the base of the CZ, r = $0.715R_\odot$. Panels from left to right correspond to the three flow profiles shown in Figure 1. The top (bottom) panels adopt the latitude-independent (latitude-dependent) pumping.

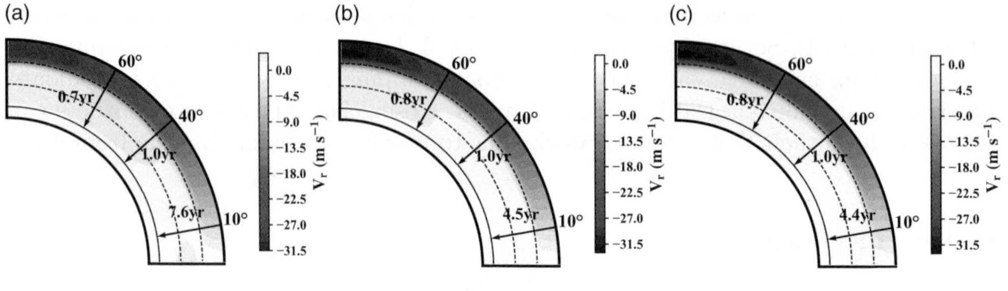

Figure 3. Inward velocity distributions. Three panels correspond to that of Figure 1 with the addition of latitude-dependent inward pumping. The black arrows indicate the transport time of the poloidal fields from the surface to the base of the CZ, at latitudes 10°, 40° and 60°.

base of CZ. Figure 2(b) and (c) show the equatorward and poleward migration of the toroidal field, respectively. Their propagational directions are determined by the flow direction at the base of CZ (see Figure 1). Then we replace the latitude-independent radial pumping by the latitude-dependent radial pumping. The corresponding results are shown in Figure 2(d)-(f) and the toroidal fields all propagate toward the equator. This means that the latitude-dependent radial flux transport is a valid mechanism for the butterfly diagram.

To illustrate how the new mechanism works, we give Figure 3 that shows the velocity distributions of the inward transport of the poloidal field. The poloidal field at higher latitudes is transported inward quicker than that at lower latitudes. Therefore, at the base the toroidal fields generated by the shearing of the poloidal field first build up at about 60° latitudes, and then build up at lower and lower latitudes.

3. Conclusions

We have proposed a new mechanism for the butterfly diagram. When considering the latitude-dependent radial transport of the poloidal field, the time- and latitude-dependent regeneration of the toroidal field causes the butterfly diagram.

Acknowledgements

The research is supported through the National Natural Science Foundation of China (grant Nos. 12173005 and 11873023).

Supplementary material

To view supplementary material for this article, please visit https://doi.org/10.1017/S1743921323000455.

References

Guerrero, G. & de Gouveia Dal Pino, E. M. 2008, A&A, 485, 267
Jiang, J., Cameron, R. H., Schmitt, D., & Işık, E. 2013, A&A, 553, A128
Pipin, V. V. & Kosovichev, A. G. 2011, ApJ Letters, 727, L45
Zhao, J., Bogart, R. S., Kosovichev, A. G., Duvall, T. L., J., & Hartlep, T. 2013, ApJ Letters, 774, L29

The Era of Multi-Messenger Solar Physics
Proceedings IAU Symposium No. 372, 2023
G. Cauzzi & A. Tritschler, eds.
doi:10.1017/S1743921323000509

Automatic detection of solar active regions from SOHO/MDI and SDO/HMI synoptic magnetograms

Ruihui Wang[1,2], Jie Jiang[1,2] and Yukun Luo[1,2]

[1]School of Space and Environment, Beihang University, Beijing, China

[2]Key Laboratory of Space Environment monitoring and Information Processing of MIIT, Beijing, China

Abstract.
We develop an adaptive method to automatically identify ARs from radial synoptic maps observed by SOHO/MDI and SDO/HMI, calibrate the detections between HMI and MDI data based on identified ARs flux and area and further derive a homogeneous dataset including ARs' area and flux over the last two solar cycles. The data are compared with sunspot number, USAF/NOAA sunspot area, SMARPs and SHARPs and BARD area and flux, which show reasonable agreement. The identified ARs during the overlap period of MDI and HMI have the same areas as a whole while the AR flux based on MDI maps is about 1.36 times as large as that of HMI maps. Based on our dataset, we find strong ARs ($|flux| > 10^{22}$Mx) contribute most to the difference between cycles 23 and 24 while other ARs ($|flux| < 10^{22}$Mx) are similar in the two cycles in both area and flux.

Keywords. Sun: activity, Sun: magnetic fields, techniques: image processing

1. Introduction

Solar active regions (ARs) play an essential role in solar physics. They not only provide insight into the solar dynamo but also lead to long and short-term solar variability. In this work, we aim to develop a new method to automatically detect ARs from SOHO/MDI and SDO/HMI radial synoptic magnetograms and to provide a new database of ARs between 1996 and 2020.

2. Method

The latitude range of detection is limited to $\pm 60°$. There are five modules in the detection algorithm. The first module is adaptive threshold segmentation to remove the background magnetic fields with different thresholds in different pixels. The second module is morphological closing operation and opening operation to remove small magnetic segments and get the ARs' kernel pixels, which are used as seeds in region growing. The third module is region growing to get all pixels of each single AR. All pixels connected to the seeds and stronger than the threshold are recovered. The fourth module is closing operation and removing decayed ARs segments that are smaller than an area threshold. The fifth module is merging neighbor regions and removing unipolar regions. We use dilation operation to merge separate segments and drop all unipolar ARs whose flux imbalance is greater than 50%.

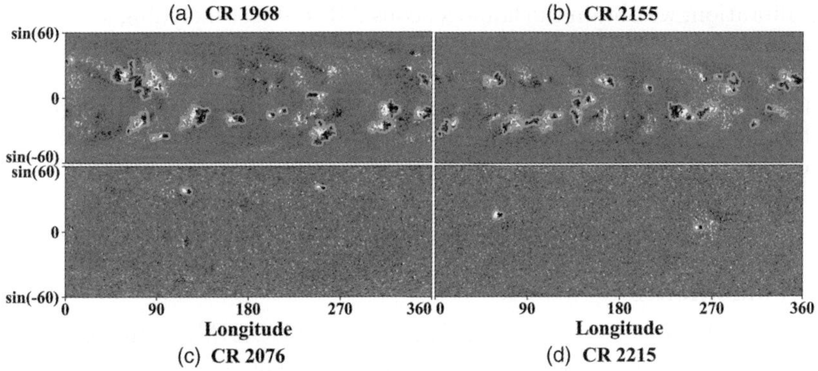

Figure 1. Examples of the detected ARs based on synoptic magnetograms. MDI (left) and HMI (right) at different phases of cycles 23 and 24 are used to demonstrate the accuracy of the algorithm. The 4 magnetograms are overplotted with the contours in dark orange outlining the perimeter of the detected ARs.

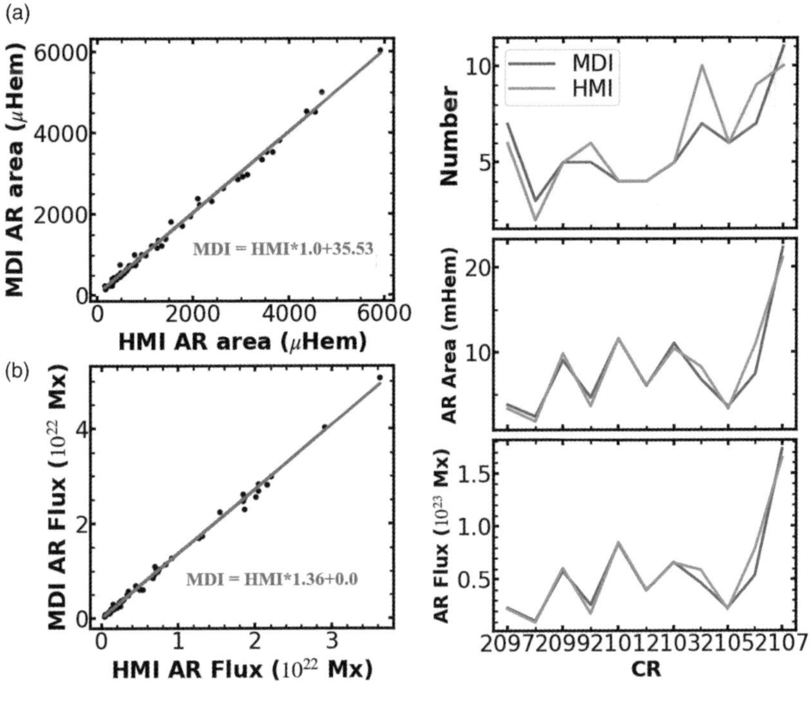

Figure 2. Calibration of the detected results based on MDI and HMI magnetograms and comparison of detected ARs during the 11 HMI and MDI overlap CRs. The ARs flux in right bottom panel is calibrated results.

3. Calibration, Validation and Statistics

Since MDI and HMI synoptic magnetograms have different resolutions, we use the magnetograms during the overlap period to calibrate the parameters used in HMI magnetograms to obtain the same ARs detections as in MDI magnetograms. The detection results during the overlap period show great agreement after calibration.

After calibration, we obtain the homogeneous ARs dataset including location, area, and flux for cycles 23 and 24. The dataset is evaluated by comparing it with other datasets, i.e., sunspot number, USAF/NOAA sunspot area, SMARPs and SHARPs (Bobra et al. (2021), Bobra et al. (2014)) area and flux, and BARD (Muñoz-Jaramillo et al. (2016)) area and flux.

We divide ARs into three categories based on their unsigned flux and find that although cycle 24 is about twice weaker than cycle 23 based on the sunspot number, the numbers and total flux of median and weak ARs in the two cycles are almost the same.

Acknowledgements

The SDO/HMI data are courtesy of NASA and the SDO/HMI team. SOHO is a project of international cooperation between ESA and NASA. The research is supported through the National Natural Science Foundation of China (grant Nos. 12173005 and 11873023).

Supplementary material

To view supplementary material for this article, please visit https://doi.org/10.1017/S1743921323000509.

References

Bobra, M. G., Sun, X., Hoeksema, J. T., et al. 2014, Sol Phys, 289, 3549

Bobra, M. G., Wright, P. J., Sun, X., & Turmon, M. J. 2021, APJS, 256, 26

Muñoz-Jaramillo, A., Werginz, Z. A., Vargas-Acosta, J. P., et al. 2016, in 2016 IEEE International Conference on Big Data (Big Data), 3194–3203

The Era of Multi-Messenger Solar Physics
Proceedings IAU Symposium No. 372, 2023
G. Cauzzi & A. Tritschler, eds.
doi:10.1017/S1743921322004975

Automatic Algorithm for Tracking Bipolar Magnetic Regions

Anu B. Sreedevi[1] and Bibhuti Kumar Jha[2]

[1]Department of Physics, Indian Institute of Technology (Banaras Hindu University), Varanasi 221005, India
email: anubsreedevi.rs.phy20@itbhu.ac.in

[2]Aryabhatta Research Institute of Observational Sciences, Nainital-263001, India

Abstract. Properties of bipolar magnetic regions (BMRs), particularly, the tilt angle play critical roles in generating the observed polar magnetic field and its reversal. Hence, a long-term study of BMR over its lifetime is crucial not only to understand the solar dynamo but also to identify the origin of the properties of BMR. In our work, we have developed an automatic algorithm to detect and track the BMRs from the line-of-sight (LOS) magnetograms of Michelson Doppler Imager (MDI) for the period of Solar Cycle 23 over its lifetime/disk passage. Here, we present the details of our algorithm and the features of BMR, particularly the tilt angle, magnetic field strength and lifetime.

Keywords. Sun: activity, Sun: magnetic fields, (Sun:) sunspots, Sun: photosphere

1. Introduction

BMRs are observed to be tilted with respect to the equator. The tilt angles are found to increase with latitude which is known as Joy's Law (Hale et al. 1919). The tilt of the BMR is vital in Babcock–Leighton mechanism, which generates the poloidal magnetic field of the Sun (Babcock 1961; Leighton 1964; Mordvinov et al. 2022). The scatter around this Joy's law produces variations in the solar cycle (Karak 2020; Karak & Miesch 2017, 2018; Lemerle & Charbonneau 2017). Stenflo & Kosovichev (2012) identified BMRs from the LOS magnetograms of MDI and found no systematic dependence between the tilt and magnetic flux of the BMR. Nevertheless, in a similar work, Jha et al. (2020) observed that for the BMRs with maximum magnetic field value greater than 2 kG, tilt decreases for increasing flux indicating tilt quenching. In the mentioned studies, every detected region is considered as a separate BMR which might affect the statistics since different BMRs will have different weightage. Here, we have tracked the identified BMRs for their lifetime or throughout their disk passage to study their properties.

2. Data and Tracking Algorithm

We use the LOS magnetogram data from MDI (Scherrer et al. 1995) and then employ a fully automatic algorithm to identify and track the BMRs. As a first step for the tracking algorithm, we identify BMRs following the method used in Jha et al. (2020), which is based on the technique explained in Stenflo & Kosovichev (2012). Our tracking algorithm is based on the idea of sunspot tracking described in Jha et al. (2021) with the needed modification in the case of BMRs.

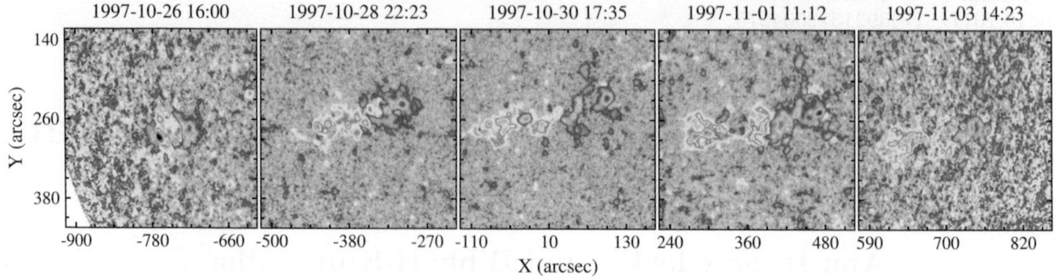

Figure 1. Snapshots of a BMR at different stage of tracking.

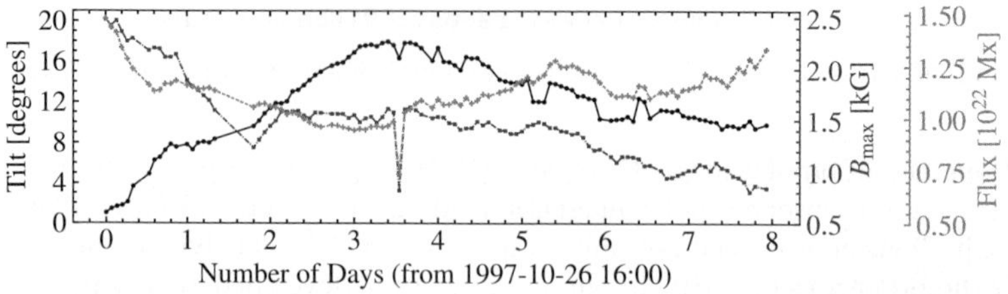

Figure 2. Evolutions of BMR tilt, flux and $B_{\rm max}$ over its lifetime.

3. Results and Conclusion

In Figure 1, we have shown one representative example of a BMR tracking algorithm. Although our algorithm has tracked this BMR from 26/10/1997 16:00:29 to 03/11/1997 14:23:44, we have demonstrated only a few intermediate tracking steps.

The analysis was done for the period of solar cycle 23. During this period, 3550 BMRs where identified and tracked. Figure 2 shows the evolution of a BMR (NOAA 8099) tracked from 26/10/1997 16:00 to 03/11/1997 14:23. Here, we observe systematic decrease of the $B_{\rm max}$ over the course of the BMRs lifetime, but the tilt is found to increase till the fourth $4^{\rm th}$ day and decrease thereafter. On studying averaged evolution of tilt over all the tracked BMRs lifetime, we find that tilt increases in the initial three days and later it remains unchanged. Whereas, $B_{\rm max}$ is found to systematically increases over the lifetime of the BMR until $6^{\rm th}$day. Detailed description of algorithm and results can be viewed at `https://youtu.be/zkc7whYhfhM` and will be published in a forthcoming article (Sreedevi et al. under preparation).

Acknowledgements

We thank Bidya Binay Karak and Dipankar Banerjee for their guidance. We also acknowledge the travel grant awarded to us by IAU to attend the conference in person.

References

Babcock, H. W. 1961, ApJ, 133, 572
Hale, G. E., Ellerman, F., Nicholson, S. B., & Joy, A. H. 1919, ApJ, 49, 153
Jha, B. K., Karak, B. B., Mandal, S., & Banerjee, D. 2020, ApJ Letters, 889, L19
Jha, B. K., Priyadarshi, A., Mandal, S., Chatterjee, S., & Banerjee, D. 2021, SoPh, 296, 25
Karak, B. B. 2020, ApJ Letters, 901, L35
Karak, B. B. & Miesch, M. 2017, ApJ, 847, 69
Karak, B. B. & Miesch, M. 2018, ApJ Letters, 860, L26
Leighton, R. B. 1964, ApJ, 140, 1547
Lemerle, A. & Charbonneau, P. 2017, ApJ, 834, 133
Mordvinov, A. V., Karak, B. B., Banerjee, D., et al. 2022, MNRAS, 510, 1331
Scherrer, P. H., Bogart, R. S., Bush, R. I., et al. 1995, SoPh, 162, 129
Stenflo, J. O. & Kosovichev, A. G. 2012, ApJ, 745, 129

Ilya Usoskin from the University of Oulu (Finland) answering questions on solar cycle reconstruction over long periods.

Chapter III:
SOLAR ATMOSPHERE AND SOLAR WIND

Aline Vidotto from Leiden University (The Netherlands) listening to questions after her talk on the evolution of the solar wind.

The Era of Multi-Messenger Solar Physics
Proceedings IAU Symposium No. 372, 2023
G. Cauzzi & A. Tritschler, eds.
doi:10.1017/S1743921322004756

How has the solar wind evolved to become what it is today?

A. A. Vidotto

Leiden Observatory, Leiden University, PO Box 9513, 2300 RA Leiden, The Netherlands

Abstract. In this contribution, I briefly review the long-term evolution of the solar wind (its mass-loss rate), including the evolution of observed properties that are intimately linked to the solar wind (rotation, magnetism and activity). I also briefly discuss implications of the evolution of the solar wind on the evolving Earth. I argue that studying exoplanetary systems could open up new avenues for progress to be made in our understanding of the evolution of the solar wind.

Keywords. stars: mass loss, stars: winds, outflows, stars: evolution

1. Introduction

The Sun is the best studied star in the whole Universe: we can measure its properties with accuracy like no other star. However, all this information just tells us about how the Sun looks like now. Understanding how the winds of cool stars evolve is key to understand, for example, how has the solar wind and the solar system planets evolved during the past 4 billion years or what the implication of a strong wind is for young exoplanets.

To understand the past, and future, evolution of the Sun, including its wind, magnetism, activity, rotation, and irradiation, we rely on data of solar-like stars, in an effort to better place the Sun and the solar wind in a stellar context. Solar-like stars form a broad group of stars with spectral types from early M to late F ($0.4 \lesssim M/M_\odot \lesssim 1.3$) in the main sequence. These stars have in common a convective outer envelope, which is linked to the generation of magnetism, which ultimately drives their high-energy radiation and winds. In this contribution, I review the evolution of some key ingredients associated to stellar winds.

2. Evolution of mass-loss rates

Detecting winds of cool dwarf stars can be very tricky, as they are in general tenuous and more difficult to detect. By modelling the observed interaction between stellar winds and the interstellar medium (ISM) Wood et al. (2002) proposed a correlation between X-ray flux and mass-loss rate. He showed that stars that have higher X-ray fluxes in general have winds with higher mass-loss rates. In addition to the method proposed by Wood et al. (2002), other methods have also been proposed to detect winds of cool dwarfs. In Vidotto (2021), I reviewed some of the methods proposed so far – this is now summarised in Figure 1, while Figure 2 † shows a compilation of these results. Following

† The values used in this plot were compiled from the following works: Drake et al. (1993); Lim et al. (1996); Gaidos et al. (2000); Wood et al. (2001, 2002, 2005, 2014); Wood & Linsky (2010); Wood (2018); Wargelin & Drake (2002); Bourrier et al. (2013); Kislyakova et al. (2014); Fichtinger et al. (2017); Vidotto & Bourrier (2017); Vidotto & Donati (2017); Jardine & Collier Cameron (2019); Finley et al. (2019); O Fionnagain et al. (2020).

How to detect winds of cool dwarfs?

Method	key requirement associated reference
Astrospheric Ly-α absorption	partially neutral ISM, ≲10 pc Wood 2004
Radio free-free emission	denser winds and/or radio flares Lim & White 1996; Güdel 1992
Using exoplanets as probes	needs evaporating planet Vidotto & Bourrier 2017
Prominences in Hα	needs fast rotation Jardine & Cameron 2019
Type II radio bursts CME-dominated	fast CME associated w/ flare Crosley & Osten 2018
Radio wave propagation	radio source (planet?) within wind Vidotto & Donati 2017
X-ray limits from stellar winds	hot coronal winds Lim & White 1996, Llama et al 2013
Charge-exchange induced X-ray	partially neutral ISM Wargelin & Drake 2001, 2002
Accretion onto white dwarfs	close binary Debes 2006

Figure 1. Table summarising (most of) the proposed methods to detect tenuous winds of cool dwarf stars (adapted from Vidotto 2021).

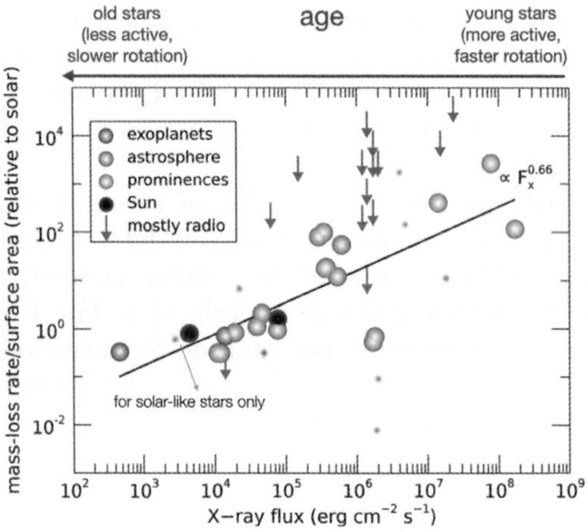

Figure 2. Relation between mass-loss rates and X-ray surface flux. Colour indicates the method used in the derivation of mass-loss rates (see panel on the left), with grey arrows indicating upper limits. The solid line is a power-law fit through the larger circles (Equation 2.1). The smaller symbols are either evolved stars or M dwarfs, which were not included in the fit and neither were the stars for which only upper limits exist (arrows). Figure from Vidotto (2021).

the suggestion from Wood et al. (2002), I present a fit for the solar-like stars (black line in Fig 2):

$$\frac{\dot{M}}{R_\star^2} = 10^{-2.75\pm0.68}\left[\frac{F_X}{\text{erg cm}^{-2}\text{s}^{-1}}\right]^{0.66\pm0.12} \frac{2\times10^{-14} M_\odot/\text{yr}}{R_\odot^2}\,. \tag{2.1}$$

In general, stars that have higher X-ray surface fluxes are younger (see Section 4). With age, stellar activity decreases. Therefore, the x-axis in Fig. 2 can also be interpreted as a roughly age indicator, whereby younger stars (which are more active and with faster rotation) are shown to the right while the older stars (which are less active and show slower rotation) are shown to the left of the plot. Using the relation from Guedel (2007), who found that $L_X \propto t^{-1.5\pm0.3}$ (for solar-like stars in the non-saturated regime) and Eq. (2.1), we have that

$$\dot{M} \propto \mathrm{age}^{-1.5\times0.66} \sim 1/\mathrm{age} \quad \text{(roughly)}\,. \tag{2.2}$$

Therefore, the mass-loss rates of solar-like stars, in the non-saturated regime, should decrease with age.

3. Stellar magnetism evolution

Because winds of cool dwarf stars are magnetic in origin (similar to the solar wind), magnetic field measurements are one of the essential input quantities in models of stellar winds†. In particular, it is interesting to note that the (large-scale) topology of the magnetic field evolves during solar cycle (e.g., DeRosa et al. 2012; Vidotto et al. 2018), which affects the observed structure of the solar wind. For example, at minimum, when the solar large-scale field can be characterised by an aligned dipole, the solar wind velocity shows a bimodal structure, with fast streams coming from the polar regions, where the solar magnetic field has open field lines (McComas et al. 2008).

To measure surface magnetic fields, I highlight two different methods. (1) Zeeman-Doppler imaging (ZDI) maps the large scale field of the stellar surface, through an image reconstruction algorithm which uses as input time-series of spectropolarimetric data. (2) By measuring the broadening of magnetically sensitive lines, the Zeeman broadening technique derives the total field strength at the stellar surface.

While the former technique is able to reconstruct the topology of the magnetic field, it can only do so on the large-scale component. While the latter technique is able to derive the total field strength (including the ZDI-undetected small-scale component), it only obtains the unsigned field (i.e., no topology indicator). As a result the two techniques are highly complementary. In depth reviews of these techniques can be found in Donati & Landstreet (2009) and Reiners (2012).

While stellar magnetism has been expected to decrease with age, it was only by grouping a relatively large sample of magnetic field observations that we finally were able to quantitatively assess this decay (Vidotto et al. 2014). Figure 3 shows that the large scale field of solar-like stars decay with $\sim \mathrm{age}^{-0.6}$, which is a rather similar slope as the "Skumanich-law", whereby rotation is observed to decay with $\sim \mathrm{age}^{-0.5}$ (Skumanich 1972). Rotational evolution will be discussed next.

4. Evolution of stellar rotation and high-energy radiation

A key concept of winds of cool dwarf stars is that they carry away angular momentum. Therefore, during the main sequence, as a star ages, its rotational velocity decreases (spin down). On the other hand, rotation is observed to be linked with (magnetic) activity, therefore, as a star spins down, its activity also decreases. Hence, older stars are characterised by slower rotation and less intense activity. There is, thus, a feedback-loop between age, rotation and (magnetic) activity, which is mediated by the angular momentum carried away by stellar winds.

Activity is a broad term that is usually related to a number of processes. Activity can be identified in the form of spots (which can be probed through photometric monitoring). Activity can also be identified through spectroscopic lines formed in the chromosphere (e.g. CaII H&K). Alternatively, it can be linked to coronal diagnostic measurements, such

† What makes solar-like stellar winds so effective at removing angular momentum is the stellar magnetic field (Weber & Davis 1967).

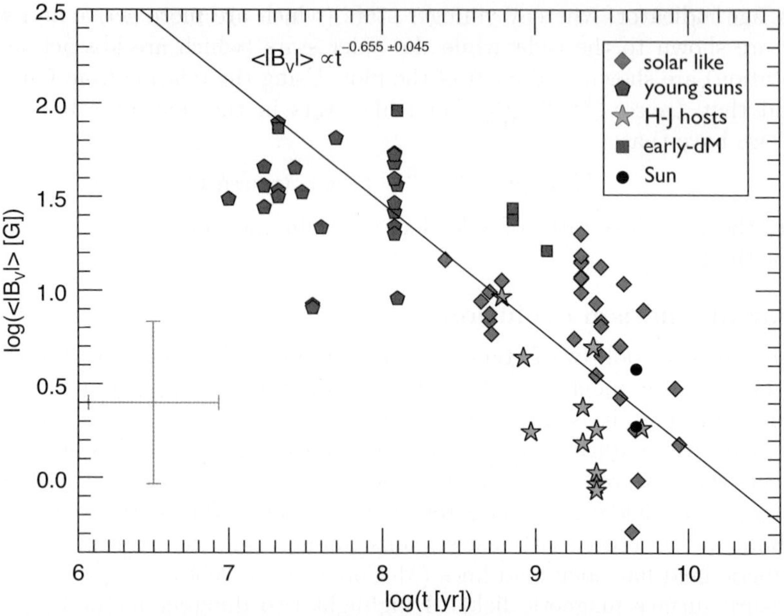

Figure 3. Evolution of the average large scale magnetic field of a sample of solar-like stars, as measured from the ZDI technique. The large scatter shown around the fit (black line) are attributed to variability of the stellar magnetic field (e.g., cycles) as well as a large uncertainty in determining stellar ages. Figure from Vidotto et al. (2014).

as X-ray emission and high-energy radiation in general. Note that all these processes (spots, chromospheric and coronal emission) are magnetic in nature, albeit they are indirect measurements of magnetism. More directly, activity can be probed by measuring stellar magnetic fields (see previous section). As activity (in the broad term) decreases with age, each of these processes are also observed to decrease with age (albeit with different slopes).

In terms of rotation, there are several observational studies dedicated to mapping the evolution of surface rotation. To reduce the uncertainty in age measurements, stars in open clusters (which are coeval) have been frequently used to to measure how stellar rotation varies as a function of age (e.g. Irwin & Bouvier 2009). In recent years, thousands of measurements of rotation periods of low-mass stars have become available thanks to new telescopes, such as Kepler (e.g. McQuillan et al. 2014). In the main sequence, stellar winds brake the rotation of stars. Therefore, models that describe stellar rotational evolution usually parameterise the evolution of angular momentum loss through stellar winds, albeit in different scales of complexity. Examples of such models are: Kawaler (1988); Spada et al. (2011); Gallet & Bouvier (2013); Johnstone et al. (2015); Matt et al. (2015); Amard et al. (2019); Pezzotti et al. (2021).

Because high-energy radiation is linked to rotation, and high-energy radiation (in particular in the ultra-violet) is very hard to measure, in recent years, semi-empirical frameworks to model and predict the evolution of high-energy radiation have been developed (Tu et al. 2015; Johnstone et al. 2021; Pezzotti et al. 2021). In general terms, these frameworks rely on models of stellar rotational evolution which are coupled to empirical relations between X-ray emission and rotation (e.g. Wright et al. 2011; Reiners et al. 2014) and to relations between UV radiation and X-ray emission (e.g. Sanz-Forcada et al. 2011; Johnstone et al. 2021). These semi-empirical frameworks provide a much needed input for models of (exo)planetary evolution, which I will discuss next.

5. Concluding remarks: effects of the evolving particle, magnetic and radiation environments of solar-like stars on (exo)planets

Understanding how the winds of cool stars evolves is key to understand wind effects on orbiting planets. A few open questions and related studies are listed below:

• *How have the solar system planets evolved during the past 4 billion years?* There are indications that Mars lost its atmosphere because of atmospheric erosion due to a stronger wind of the young sun (Kulikov et al. 2007). Earth, on the other hand, is believed to have retained its atmosphere because of its strong intrinsic magnetic field (Blackman & Tarduno 2018), in spite of the fact that its magnetosphere would have been smaller at earlier ages (Carolan et al. 2019).

• *What is the implication of a strong wind on atmospheric evaporation of close-in exoplanets?* The wind environment surrounding close-in exoplanets are harsher than the environment around planets orbiting far from their host stars (Vidotto et al. 2015). For unmagnetised planets, this could enhance their atmospheric evaporation (McCann et al. 2019; Hazra et al. 2022, but see also Vidotto & Cleary 2020; Carolan et al. 2020b). The interaction with a stellar wind can also shape the atmospheric material escaping from the planet (Bourrier & Lecavelier des Etangs 2013; Debrecht et al. 2019; Villarreal D'Angelo et al. 2021; Kubyshkina et al. 2022). Even if the effect of winds on erosion is not significant, it can substantially affect their observational signatures of atmospheric evaporation through spectroscopic transits (Carolan et al. 2020a,b).

• *How strong are the interactions between a close-in planet and the wind of its planet-hosting star?* There are different types of star-planet interactions (Vidotto 2020), including interactions that are mediated by the stellar wind, such as planetary auroral radio emission (Kavanagh et al. 2019; Turner et al. 2021) or planet-induced emission on the host star (Vedantham et al. 2020; Kavanagh et al. 2021). The strength of these interactions depend on the local (i.e., at the planet's orbit) conditions of the stellar wind, which are higher for planets orbiting closer to their host star. For this reason, star-planet interactions mediated by the wind particles and its magnetic fields are stronger in closer-in planets (Strugarek 2016).

• *Are stellar winds big "cosmic ray filters"?* Cosmic rays are ionised, high-energy particles that permeate the Galaxy (in which case they are known as galactic cosmic rays) or particles that originate in stellar flares or coronal mass ejections (in which case they are know as SEPs or stellar energetic particles). Once these particles are inside the stellar wind, the magnetised wind can modulate them both in momentum and spatial domains (what I refer above as giant cosmic ray filters). In essence, the wind can push cosmic rays out, as they are spatially advected, or can allow them in, as they diffuse into the system (e.g., Potgieter 2013). These processes depend on the properties of stellar winds, which evolve in time (Cohen et al. 2012; Rodgers-Lee et al. 2020, 2021a,b). Knowing whether cosmic rays have been more or less suppressed in the past† are important for chemical modelling of planetary atmospheres: cosmic rays can penetrate atmospheres, providing an extra source of atmospheric ionisation, driving the formation of prebiotic molecules, and dissociating molecules (e.g. Rimmer & Helling 2013; Atri 2017; Barth et al. 2021).

The strength of some star-planet interactions (e.g., planetary auroral emission, the interplay between an evaporating atmosphere and the stellar wind, cosmic ray suppression) depends directly on the local and/or global characteristics of the stellar wind. Therefore, an interesting "by-product" of studying star-planet interactions, in particular interactions with the stellar wind, is that if such interactions can be observed, they can

† Also around other planetary systems (e.g., Fraschetti et al. 2019; Herbst et al. 2020; Mesquita et al. 2021, 2022a,b).

help characterise not only the planet but also the stellar wind of cool dwarf stars, which form the majority of planet hosts known nowadays.

Acknowledgements

I thank the SOC of the IAUS372 for the invitation to participate in this symposium. In particular, I thank the SOC chairs, Gianna Cauzzi and Ali Tritschler, for their impeccable organisation of this meeting during very difficult pandemic times. AAV has received funding from the European Research Council (ERC) under the European Union's Horizon 2020 research and innovation programme (grant agreement No 817540, ASTROFLOW).

References

Amard, L., Palacios, A., Charbonnel, C., et al. 2019, A&A, 631, A77
Atri, D. 2017, MNRAS, 465, L34
Barth, P., Helling, C., Stüeken, E. E., et al. 2021, MNRAS, 502, 6201
Blackman, E. G. & Tarduno, J. A. 2018, MNRAS, 481, 5146
Bourrier, V. & Lecavelier des Etangs, A. 2013, A&A, 557, A124
Bourrier, V., Lecavelier des Etangs, A., Dupuy, H., et al. 2013, A&A, 551, A63
Carolan, S., Vidotto, A. A., Loesch, C., & Coogan, P. 2019, MNRAS, 489, 5784
Carolan, S., Vidotto, A. A., Plavchan, P., D'Angelo, C. V., & Hazra, G. 2020a, MNRAS, 498, L53
Carolan, S., Vidotto, A. A., Villarreal D'Angelo, C., & Hazra, G. 2020b, MNRAS, 500, 3382
Cohen, O., Drake, J. J., & Kota, J. 2012, ApJ, 760, 85
Debrecht, A., Carroll-Nellenback, J., Frank, A., et al. 2019, MNRAS, 483, 1481
DeRosa, M. L., Brun, A. S., & Hoeksema, J. T. 2012, ApJ, 757, 96
Donati, J. & Landstreet, J. D. 2009, Annual Review of Astronomy & Astrophysics, 47, 333
Drake, S. A., Simon, T., & Brown, A. 1993, ApJ, 406, 247
Fichtinger, B., Guedel, M., Mutel, R. L., et al. 2017, A&A, 599, A127
Finley, A. J., Hewitt, A. L., Matt, S. P., et al. 2019, ApJ Letters, 885, L30
Fraschetti, F., Drake, J. J., Alvarado-Gomez, J. D., et al. 2019, ApJ, 874, 21
Gaidos, E. J., Guedel, M., & Blake, G. A. 2000, Geophysical Research Letters, 27, 501
Gallet, F. & Bouvier, J. 2013, A&A, 556, A36
Guedel, M. 2007, Living Reviews in Solar Physics, 4, 3
Hazra, G., Vidotto, A. A., Carolan, S., Villarreal D'Angelo, C., & Manchester, W. 2022, MNRAS, 509, 5858
Herbst, K., Scherer, K., Ferreira, S. E. S., et al. 2020, ApJ Letters, 897, L27
Irwin, J. & Bouvier, J. 2009, in IAU Symposium, Vol. 258, IAU Symposium, ed. E. E. Mamajek, D. R. Soderblom, & R. F. G. Wyse, 363–374
Jardine, M. & Collier Cameron, A. 2019, MNRAS, 482, 2853
Johnstone, C. P., Bartel, M., & Güdel, M. 2021, A&A, 649, A96
Johnstone, C. P., Guedel, M., Brott, I., & Lueftinger, T. 2015, A&A, 577, A28
Kavanagh, R. D., Vidotto, A. A., Klein, B., et al. 2021, MNRAS, 504, 1511
Kavanagh, R. D., Vidotto, A. A., O Fionnagain, D., et al. 2019, MNRAS, 485, 4529
Kawaler, S. D. 1988, ApJ, 333, 236
Kislyakova, K. G., Holmstrom, M., Lammer, H., Odert, P., & Khodachenko, M. L. 2014, Science, 346, 981
Kubyshkina, D., Vidotto, A. A., Villarreal D'Angelo, C., et al. 2022, MNRAS, 510, 2111
Kulikov, Y. N., Lammer, H., Lichtenegger, H. I. M., et al. 2007, Space Science Review, 129, 207
Lim, J., White, S. M., & Slee, O. B. 1996, ApJ, 460, 976
Matt, S. P., Brun, A. S., Baraffe, I., Bouvier, J., & Chabrier, G. 2015, ApJ Letters, 799, L23
McCann, J., Murray-Clay, R. A., Kratter, K., & Krumholz, M. R. 2019, ApJ, 873, 89
McComas, D. J., Ebert, R. W., Elliott, H. A., et al. 2008, Geophysical Research Letters, 35, 18103
McQuillan, A., Mazeh, T., & Aigrain, S. 2014, ApJ Supp., 211, 24

Mesquita, A. L., Rodgers-Lee, D., & Vidotto, A. A. 2021, MNRAS, 505, 1817
Mesquita, A. L., Rodgers-Lee, D., Vidotto, A. A., Atri, D., & Wood, B. E. 2022a, MNRAS, 509, 2091
Mesquita, A. L., Rodgers-Lee, D., Vidotto, A. A., & Kavanagh, R. D. 2022b, MNRAS, 515, 1218
O Fionnagain, D., Vidotto, A. A., Petit, P., et al. 2020, MNRAS, 500, 3438
Pezzotti, C., Attia, O., Eggenberger, P., Buldgen, G., & Bourrier, V. 2021, A&A, 654, L5
Potgieter, M. S. 2013, Living Reviews in Solar Physics, 10, 3
Reiners, A. 2012, Living Reviews in Solar Physics, 9, 1
Reiners, A., Schuessler, M., & Passegger, V. M. 2014, ApJ, 794, 144
Rimmer, P. B. & Helling, C. 2013, ApJ, 774, 108
Rodgers-Lee, D., Taylor, A. M., Vidotto, A. A., & Downes, T. P. 2021a, MNRAS, 504, 1519
Rodgers-Lee, D., Vidotto, A. A., & Mesquita, A. L. 2021b, MNRAS, 508, 4696
Rodgers-Lee, D., Vidotto, A. A., Taylor, A. M., Rimmer, P. B., & Downes, T. P. 2020, MNRAS, 499, 2124
Sanz-Forcada, J., Micela, G., Ribas, I., et al. 2011, A&A, 532, A6
Skumanich, A. 1972, ApJ, 171, 565
Spada, F., Lanzafame, A. C., Lanza, A. F., Messina, S., & Collier Cameron, A. 2011, MNRAS, 416, 447
Strugarek, A. 2016, ApJ, 833, 140
Tu, L., Johnstone, C. P., Guedel, M., & Lammer, H. 2015, A&A, 577, L3
Turner, J. D., Zarka, P., Grießmeier, J.-M., et al. 2021, A&A, 645, A59
Vedantham, H. K., Callingham, J. R., Shimwell, T. W., et al. 2020, Nature Astronomy, 4, 577
Vidotto, A. A. 2020, IAU Symposium, 354, 259
Vidotto, A. A. 2021, Living Reviews in Solar Physics, 18, 3
Vidotto, A. A. & Bourrier, V. 2017, MNRAS, 470, 4026
Vidotto, A. A. & Cleary, A. 2020, MNRAS, 494, 2417
Vidotto, A. A. & Donati, J.-F. 2017, A&A, 602, A39
Vidotto, A. A., Fares, R., Jardine, M., Moutou, C., & Donati, J.-F. 2015, MNRAS, 449, 4117
Vidotto, A. A., Gregory, S. G., Jardine, M., et al. 2014, MNRAS, 441, 2361
Vidotto, A. A., Lehmann, L. T., Jardine, M., & Pevtsov, A. A. 2018, MNRAS, 480, 477
Villarreal D'Angelo, C., Vidotto, A. A., Esquivel, A., Hazra, G., & Youngblood, A. 2021, MNRAS, 501, 4383
Wargelin, B. J. & Drake, J. J. 2002, ApJ, 578, 503
Weber, E. J. & Davis, L. J. 1967, ApJ, 148, 217
Wood, B. E. 2018, in Journal of Physics Conference Series, Vol. 1100, Journal of Physics Conference Series, 012028
Wood, B. E. & Linsky, J. L. 2010, ApJ, 717, 1279
Wood, B. E., Linsky, J. L., Mueller, H., & Zank, G. P. 2001, ApJ Letters, 547, L49
Wood, B. E., Mueller, H.-R., Redfield, S., & Edelman, E. 2014, ApJ Letters, 781, L33
Wood, B. E., Mueller, H.-R., Zank, G. P., & Linsky, J. L. 2002, ApJ, 574, 412
Wood, B. E., Mueller, H.-R., Zank, G. P., Linsky, J. L., & Redfield, S. 2005, ApJ Letters, 628, L143
Wright, N. J., Drake, J. J., Mamajek, E. E., & Henry, G. W. 2011, ApJ, 743, 48

The Era of Multi-Messenger Solar Physics
Proceedings IAU Symposium No. 372, 2023
G. Cauzzi & A. Tritschler, eds.
doi:10.1017/S1743921322004744

Spectroscopic Detection of Alfvénic MHD Waves in the Sunspot Chromosphere

Jongchul Chae

Seoul National University, Seoul, Korea

Abstract. Alfvénic waves are regarded as an important process in understanding coronal heating, solar wind acceleration, and the fractionization of low first-ionization-potential (FIP) elements. Recently, significant progresses have been made in the detection of propagating Alfvénic waves in the solar chromosphere using two different methods: the imaging method and the spectroscopic method. The imaging method detects Alfvénic waves that oscillate in the direction perpendicular to the line of sight, and the spectroscopic method, those that oscillates in the line of sight direction. We have applied the spectroscopic method to the imaging spectral data taken by the FISS on GST at Big Bear. As a result, we detected a number of propagating Alfvénic wave packets, and found that there are two distinct groups: three-minute period waves, and ten-minute period waves.

Keywords. line: profiles, MHD, waves, techniques: spectroscopic, Sun: atmospheric motions, Sun: chromosphere, Sun: oscillations, sunspots

1. Introduction

Alfvénic MHD waves refer to transverse MHD waves that propagate along magnetic field lines. They include pure Alfvén waves, either shear Alfvén waves in a planar geometry or torsional Alfvén waves in a magnetic flux tube, that are fully incompressive, and kink waves in a magnetic flux tube, that are weakly compressive. Even though pure Alfvén waves and kink waves differ in their physical nature (Van Doorsselaere et al. 2008), they are similar to each other in that both are transverse and both propagate along magnetic field lines. Moreover, they can be mutually converted under some conditions (Cranmer & van Ballegooijen 2005), which justifies the use of the term Alfvénic waves without having to discern between the two kinds.

Alfvénic waves have been theoretically investigated for long time because they are considered important in the solar research. As they are an efficient carrier of energy, they are believed important in the problem of coronal heating (Alfvén 1947; Osterbrock 1961). In addition, Alfvénic waves produce the pondermotive force that can drive charged particles along magnetic field lines (Litwin & Rosner 1998), so they may hold keys to understanding the problem of solar wind acceleration (Cranmer & van Ballegooijen 2005), and the problem coronal abundance enhancement of low FIP elements (Laming 2004).

The detection of Alfvénic waves has been mostly done with the imaging method. In this method, transverse kink motions on the plane of sky are inferred from the time variation of the shape of chromospheric threads. This method was successfully used to detect Alfvénic waves in the limb spicules (De Pontieu et al. 2007), and disk fibrils either in active regions (He et al. 2009; Pietarila et al. 2011) or in quiet regions (Jess et al. 2012). It was also shown with this method that Alfvénic waves pervade in sunspot superpenumbral fibrils (Morton et al. 2021).

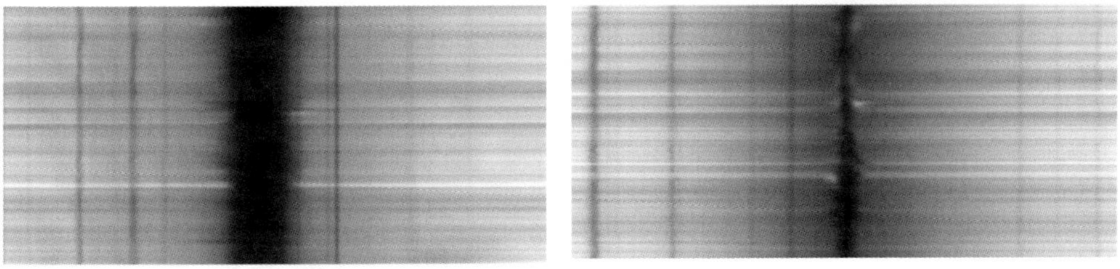

Figure 1. Example of the Hα spectrogram (left) and the Ca II 8542 spectrogram (right) simultaneously taken by the FISS.

The spectroscopic detection of Alfvénic waves in the chromosphere has been very rare. A very earlier attempt to determine the line-of-sight velocity oscillations associated with Alfvénic waves in chromospheric fibrils was made by Giovanelli (1975) based on the spectral scan of the Hα line using a tunable narrowband birefringent filter. As far as we know, we are the first who clearly detected Alfvénic waves in chromospheric fibrils with a real spectrograph. In this proceeding, we will briefly review the spectroscopic data and method we have used, and the main findings we have obtained (Chae et al. 2021b, 2022).

2. Data and Method

To obtain the spectral data, we used the Fast Imaging Solar Spectrograph (FISS) on the 1.6 meter Goode Solar Telescope at the Big Bear Solar Observatory. The FISS is a dual-band Echelle Spectrograph with the imaging capability based on the raster scan of the slit over the field of view (Chae et al. 2013). By default, it simultaneously records the spectrogram around the Hα line by camera A and the spectrogram around the Ca II 8542 line by camera B (see Fig 1). The spectral resolving power is higher than 10^5 in both bands. The raster scan of a $20''$-width field of view is typically done every 20 s. The spatial resolution depends on the seeing and the performance of the instrument, and can be as high as $0.''32$.

The physical parameters are inferred from each line profile taken at a point at a specific instant based on spectral inversion. For this purpose, we have developed the technique of multilayer spectral inversion (MLSI, Chae et al. 2020, 2021a). In this technique, the atmospheric consists of three layers: the photosphere, the lower chromosphere, and the upper chromosphere. The radiative transfer in each layer is described by the parameters: source function, Doppler width, and line-of-sight velocity, with their height variations being described by the first polynomials of optical depth. The model has ten free parameters, seven among which are the parameters of the chromospheric layers. The left panel of Fig. 2 illustrates the MLSI fitting of the Hα line profile.

The most useful parameters are the line-of-sight velocity, the Doppler width, and the source function at the upper layer where the line core is formed. Particularly the line-of-sight velocity data is used to detect velocity oscillations along the line of sight. Once the inversion is done at every point inside the field of view, one can construct the map of line-of-sight velocity as shown in the right panel of Fig. 2. The time series of spatially aligned velocity maps allows us to investigate the velocity oscillation at an arbitrary point inside the field of view.

The principle of spectroscopic Alfvénic wave detection is rather simple. We choose chromospheric features like fibrils that are highly elongated in the plane of sky. In a feature of this kind, magnetic field lines are likely to be quite along the direction of elongation direction, and to be very much perpendicular to the line of sight. Then, as illustrated in Fig. 3, the oscillations of line-of-sight velocity at a point on the feature

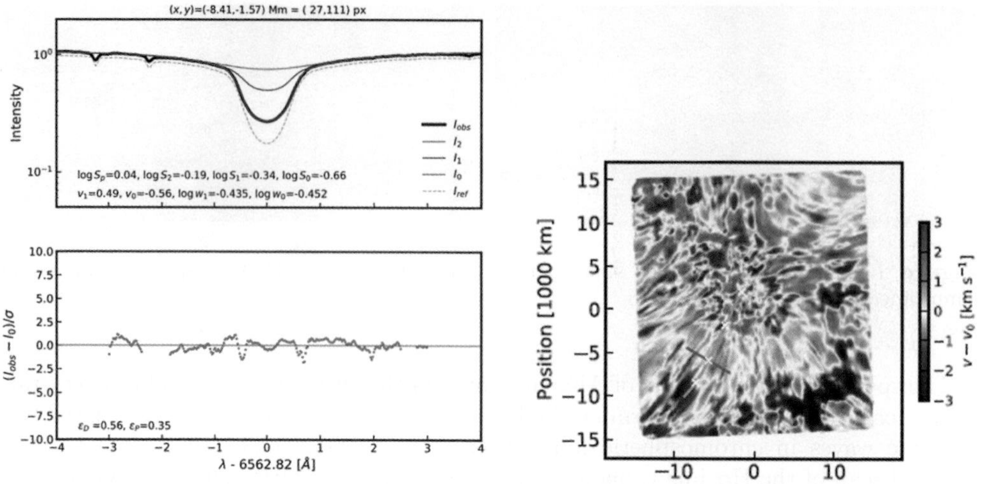

Figure 2. Example of the MLSI fitting of the Hα line profile (left) and the map of the line-of-sight velocity inferred from the Hα line inversion (right).

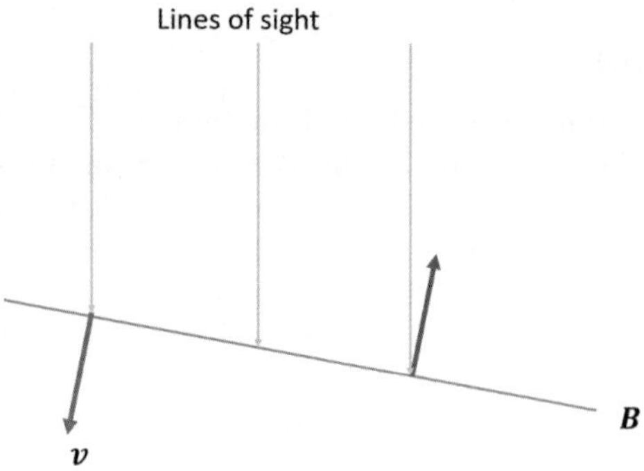

Figure 3. Illustration of spectroscopic Alfvén wave detection.

is mostly attributed to the oscillations of transverse velocity across the field lines. The analysis of these oscillations will yield the value of velocity amplitude as well as wave period. Moreover, if the velocity oscillations at two points that are located apart in the same feature are correlated with each other with a finite time lag, one can determine the direction and speed of wave propagation along the feature, which is expected to be comparable to the Alfvén speed in the chromosphere.

Our data provide us with another test for the verification of Alfvén waves. Since Alfvén waves are either incompressive or weakly compressive, the velocity oscillations will not accompany the oscillations of pressure and temperature. Temperature can be determined from the Doppler width of the Hα line. We can check whether velocity oscillations and temperature oscillations are strongly correlated with each other or not.

3. Results

The first spectroscopic detection of Alfvénic waves in the chromosphere was done by Chae et al. (2021b) in numerous fibrils around a small sunspot. The Hα intensity map of the observed region is shown in the upper left of Fig. 4. Point A corresponds to the center of this sunspot. As illustrated in the upper right of the figure, the line of sight at this point is almost parallel to the magnetic field lines, so the line-of-sight

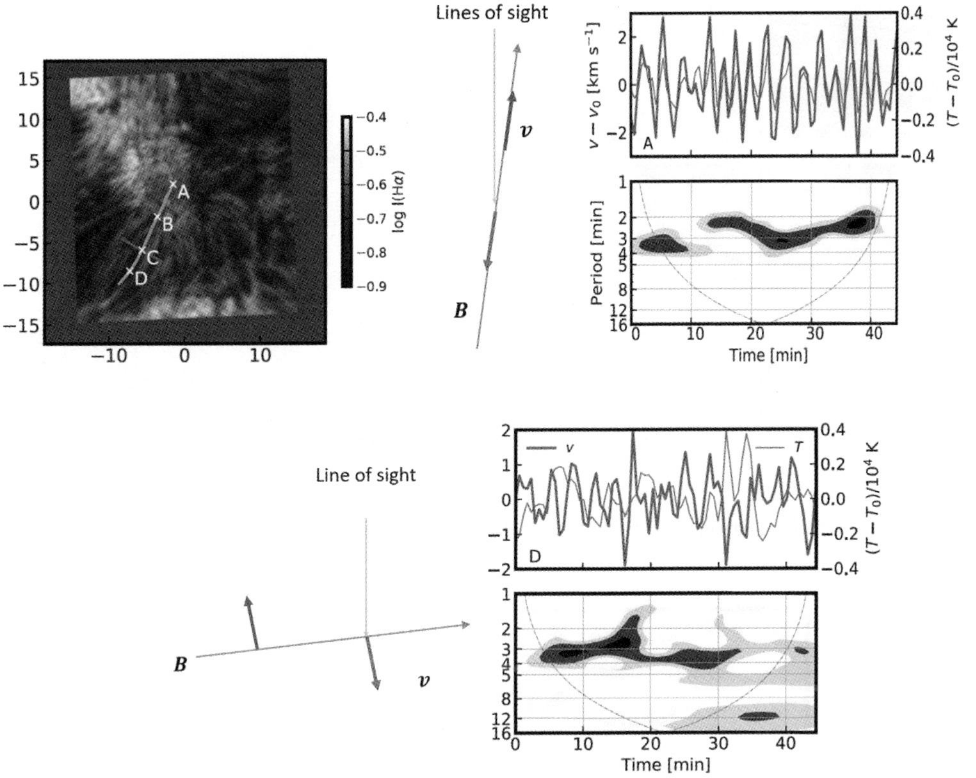

Figure 4. Points of interest marked on the map of Hα core intensity (upper left), the oscillations of velocity and temperature at point A (upper right), and those at point D (lower).

velocity oscillations represent the longitudinal velocity oscillations. These are the well-known umbral oscillations of velocity that are interpreted as slow MHD waves under the influence of gravity. At this point A, the temperature fluctuations well correlated with the velocity oscillations, which is consistent with the compressive nature of the waves.

Point D is located on a fibril emanating from the sunspot. At this point, magnetic field lines are predominantly perpendicular to the line of sight as depicted at the bottom of Fig. 4. Therefore, the velocity oscillations measured at this point shown at the bottom of Fig. 4. are likely to represent mostly the transverse oscillations of velocity. The figure also presents the temperature fluctuations at D, which are not correlated with the velocity oscillations at all, suggesting the incompressive nature of the velocity oscillations. The time-distance map of velocity constructed along the fibril (shown in Chae et al. 2021b) indicates that these velocity oscillations propagate at speeds around 100 km s^{-1}. Therefore, the velocity oscillations at D may be attributed to Alfvénic waves in the chromospheric fibril.

It was later realized by Chae et al. (2022) that the detection of Alfvénic waves of different periods can be done more systematically by decomposing the velocity fluctuations into a few band-filtered oscillations. Fig. 5 presents some maps of band-filtered velocity as well as that of the non-filtered velocity. The map of velocity with periods from 3.0 to 7.0 min reveals red-colored (i.e. redshifted) velocity stripes that are radially extended, much more clearly than the map of non-filtered velocity. The velocity stripes of this kind is a signature of the presence of Alfvénic waves because Alfvénic waves have a high speed of propagation and a very long wavelength. The value of propagation speed

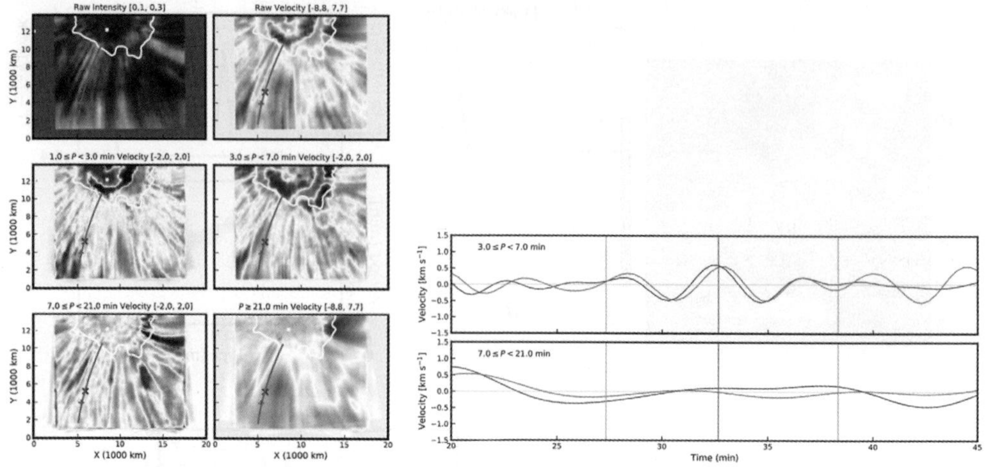

Figure 5. Maps of Hα core intensity, non-filtered line-of-sight velocity, and band-filtered velocities (left), and illustration of the cross-correlation analysis of band-filtered velocity oscillations at two points (right).

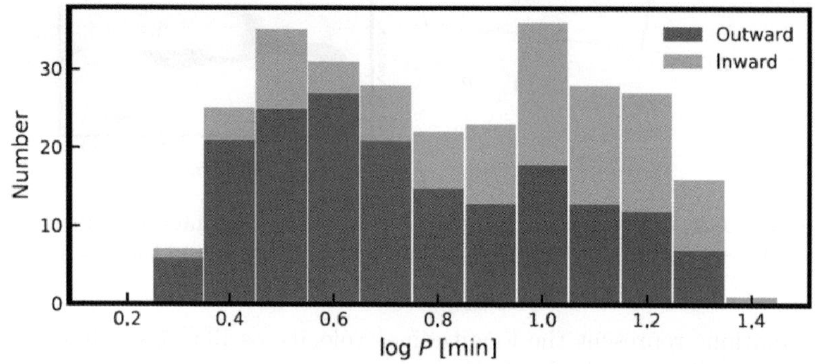

Figure 6. Number distribution of wave period indicating two groups of waves: three-minute waves and ten-minute waves.

can be determined by the analysis of cross correlation between the band-filtered velocity oscillations at two points as illustrated at the right of Fig. 5. In this specific case, the propagation speed was found to be 80 km s^{-1}.

We detected Alfvénic waves 279 times at different locations and at different times. From each detection, we determined the values of period, velocity amplitude, and propagation speed as well as propagation direction. The most interesting is the number distribution of wave period as shown in Fig. 6. The Alfvénic waves we detected can be divided into two groups depending on the period: the group of waves (three-minute waves) with periods around $10^{0.5}$ min and another (ten-minute waves) with periods around $10^{1.0}$ min. The figure also indicates that the three-minutes waves propagate mostly outward from the sunspot, but the ten-minute waves propagate either outward from or inward to the sunspot with roughly equal occurrence rates. Summarized in Table 1 are the typical values of period, propagation speed, and velocity amplitude in the two groups of waves.

4. Summary and Conclusion

From the imaging spectrograph data taken with the FISS on the GST at the Big Bear Solar Observatory, we have successfully constructed the time series of line-of-sight

Table 1. Typical values of observed parameters in three-minute waves and ten-minute waves.

Parameter	Three-minute waves	Ten-minute waves
Period	3.7 min	12 min
Propagation speed	140 km s^{-1}	65 km s^{-1}
Velocity amplitude	0.7 km s^{-1}	0.6 km s^{-1}

velocity maps. Using these velocity data, we successfully detected Alfvénic waves in the chromospheric fibrils surrounding sunspots that oscillate in the vertical direction. The absence of intensity-velocity correlation and the values of determined propagation speeds comparable to Alfvén speeds in the chromosphere strongly support the detection of Alfvénic waves. Our results also strongly suggest that there are two types of Alfvénic waves: three-minute waves and ten-minute waves.

Our study opens the door to the further spectroscopic investigation of Alfvénic waves in the solar chromosphere, not only on sunspot regions, but also in plage regions and quiet regions. It also raises a number of questions that have to be resolved in the future. From the theoretician's point of view, what are the origin of three-minute waves? Is the slow-to-Alfvénic mode conversion we proposed (Chae et al. 2021b) really operating or not? How about ten-minute wave? Are they related to the cutoff period of Alfvénic wave as we proposed (Chae et al. 2022) or not? Do the Alfvénic wave detected with the spectroscopic method much overlap those detected with the imaging method or do they represent distinct kinds of Alvénic waves? From the observer's point of view, is there any way of cleanly separating between compressive slow waves and incompressive Alfvénic waves in penumbral regions as it is likely that these two kinds exist together. If possible, the study of sunspot waves will be very fruitful.

Acknowledgements

This research was supported by the National Research Foundation of Korea (NRF-2020R1A2C2004616 and NRF-2020R1I1A1A01068789). The GST operation is partly supported by the Korea Astronomy and Space Science Institute, the Seoul National University, and the Key Laboratory of Solar Activities of Chinese Academy of Sciences (CAS) and the Operation, Maintenance and Upgrading Fund of CAS for Astronomical Telescopes and Facility Instruments.

References

Alfvén, H. 1947, MNRAS, 107, 211
Chae, J., Cho, K., Kang, J., et al. 2021a, Journal of Korean Astronomical Society, 54, 139
Chae, J., Cho, K., Lim, E.-K., & Kang, J. 2022, ApJ, 933, 108
Chae, J., Cho, K., Nakariakov, V. M., Cho, K.-S., & Kwon, R.-Y. 2021b, ApJ Letters, 914, L16
Chae, J., Madjarska, M. S., Kwak, H., & Cho, K. 2020, A&A, 640, A45
Chae, J., Park, H.-M., Ahn, K., et al. 2013, Solar Phys., 288, 1
Cranmer, S. R. & van Ballegooijen, A. A. 2005, ApJ Suppl., 156, 265
De Pontieu, B., McIntosh, S. W., Carlsson, M., et al. 2007, Science, 318, 1574
Giovanelli, R. 1975, Solar Phys., 44, 299
He, J. S., Tu, C. Y., Marsch, E., et al. 2009, A&A, 497, 525
Jess, D. B., Pascoe, D. J., Christian, D. J., et al. 2012, ApJ Letters, 744, L5
Laming, J. M. 2004, ApJ, 614, 1063
Litwin, C. & Rosner, R. 1998, ApJ Letters, 506, L143
Morton, R. J., Mooroogen, K., & Henriques, V. M. J. 2021, Philosophical Transactions of the Royal Society of London Series A, 379, 20200183
Osterbrock, D. E. 1961, ApJ, 134, 347
Pietarila, A., Aznar Cuadrado, R., Hirzberger, J., & Solanki, S. K. 2011, ApJ, 739, 92
Van Doorsselaere, T., Nakariakov, V. M., & Verwichte, E. 2008, ApJ Letters, 676, L73

The Era of Multi-Messenger Solar Physics
Proceedings IAU Symposium No. 372, 2023
G. Cauzzi & A. Tritschler, eds.
doi:10.1017/S1743921322004690

Self-consistent nanoflare heating in model active regions: MHD avalanches in curved coronal arcades

J. Reid[1], J. Threlfall[2] and A. W. Hood[1]

[1]University of St Andrews, St Andrews, United Kingdom

[2]Abertay University, Dundee, United Kingdom

Abstract. MHD avalanches involve small, narrowly localized instabilities spreading across neighbouring areas in a magnetic field. Cumulatively, many small events release vast amounts of stored energy. Straight cylindrical flux tubes are easily modelled, between two parallel planes, and can support such an avalanche: one unstable flux tube causes instability to proliferate, via magnetic reconnection, and then an ongoing chain of like events. True coronal loops, however, are visibly curved, between footpoints on the same solar surface. With 3D MHD simulations, we verify the viability of MHD avalanches in the more physically realistic, curved geometry of a coronal arcade. MHD avalanches thus amplify instability across strong solar magnetic fields and disturb wide regions of plasma. Contrasting with the behaviour of straight cylindrical models, a modified ideal MHD kink mode occurs, more readily and preferentially upwards in the new, curved geometry. Instability spreads over a region far wider than the original flux tubes and than their footpoints. Consequently, sustained heating is produced in a series of 'nanoflares' collectively contributing substantially to coronal heating. Overwhelmingly, viscous heating dominates, generated in shocks and jets produced by individual small events. Reconnection is not the greatest contributor to heating, but is rather the facilitator of those processes that are. Localized and impulsive, heating shows no strong spatial preference, except a modest bias away from footpoints, towards the loop's apex. Remarkable evidence emerges of 'campfire' like events, with simultaneous, reconnection-induced nanoflares at separate sites along coronal strands, akin to recent results from Solar Orbiter. Effects of physically realistic plasma parameters, and the implications for thermodynamic models, with energetic transport, are discussed.

Keywords. Sun: corona, Sun: magnetic fields, magnetohydrodynamics, methods: numerical

1. Introduction

One of the most long-standing open questions in solar physics is the coronal heating problem, namely why the Sun's temperature should rise further away from its surface. Heating is greatest in coronal loops that visibly curve between footpoints on the photosphere. Several explanations are offered, including MHD 'avalanches'. Some natural systems, such as the coronal magnetic field, are posited to be self-organized and critical, such that one local event can dissipate a small amount of energy and trigger further such small events at stressed, yet stable, neighbouring sites. A chain reaction—an 'avalanche'—occurs, releasing great energy through the cumulative action of self-similar events. Convective motions inject energy into the coronal field. When sufficiently stressed, magnetic reconnection reconfigures, and releases energy from, the field (Lu & Hamilton 1991). At one extreme, these events trigger flares; at the other, they are scarcely detectable, giving weak, underlying heating: 'nanoflares' (Parker 1988).

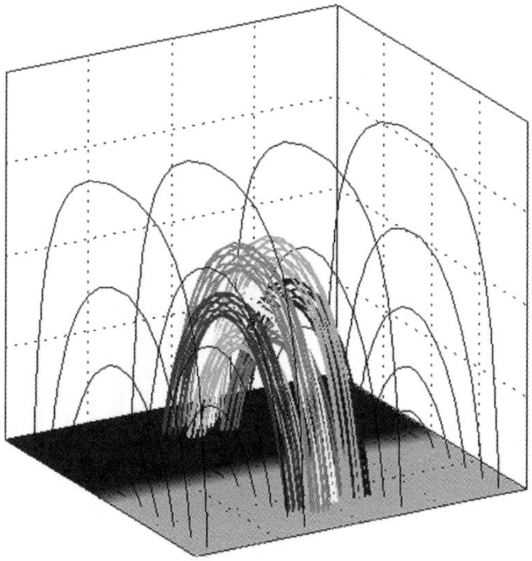

Figure 1. Initial arcade.

Loops are typically modelled as straightened cylinders, between two parallel planes, which are far simpler to conceive, study, and simulate. Recently, truly curved models have become more common. Can, therefore, the first type of model accurately reproduce the behaviour of the more realistic second? One can also compare the effect on heating.

2. Model

A curved magnetic field is modelled, with field lines anchored at footpoints on the same plane, as in bipolar arcades (Figure 1). Vortical motions twist these footpoints, creating seven magnetic flux tubes. *Lare3d* (Arber et al. 2001), an MHD code, advances the model in time. Shock heating mimics the physics of shocks, and background viscosity reflects the general viscous damping. Anomalous resistivity dissipates strong currents.

3. Results & conclusions

Twisting motions create helical current sheets. In the fastest-twisted central tube, this appears as a thin crescent in a cross-section, intense enough to cause an ideal MHD instability, which enables resistivity, starts an unstable, dissipative phase, and disrupts the first tube, which progressively disturbs each surrounding it. Substantial heating arises in a series of 'bursts' atop a fairly steady background (Figure 2). Each such event corresponds to a sharp fall in magnetic energy, growth in kinetic energy via fast outflows, and consequently higher thermal energy from viscosity and Ohmic dissipation.

Total heating mostly consists of shock heating and viscosity, with little Ohmic heating (Figure 2). In one heating event, disruption begins with strong currents causing a magnetic instability and thereby reconnection in a small diffusion region, yielding Ohmic heating, which thus precedes other forms. Strong outflows and jets are then accelerated, which may form shocks and contribute to turbulence in the plasma. Viscous heating then surpasses Ohmic. Spatial distributions of heating and current show the same property. Currents are strong in fine, field-aligned layers, but heating is smoother and more diffuse, generated by fluid and plasma processes, such as shocks, jets, and turbulence, around narrow reconnection regions.

MHD avalanches are shown equally viable in curved and straightened models, vindicating the simplified models. Heating in a curved arcade is dispersed, highly localized,

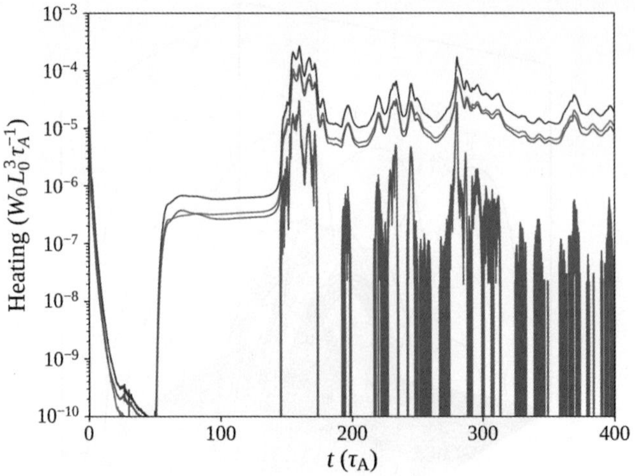

Figure 2. Total heating (blue) composed of shock (red), viscous (magenta), and Ohmic (green).

and lacking spatial preference. Magnetic reconnection is not itself a major source heating, but rather facilitates the processes, such as shocks, jets, and turbulence, that are.

Enhanced computational resources and refined numerical methods are now enabling a more detailed study of the aftermath and effect of the heating. Thermodynamic properties, even around the steep gradients in temperature in the transition region, will be resolvable in MHD, and thus the thermal response of the plasma to heating within reach.

References

Arber, T. D., Longbottom, A. W., Gerrard, C. L., & Milne, A. M. 2001, J. Comput. Phys, 171, 151

Lu, E. T. & Hamilton, R. J. 1991, ApJ, 380, L89

Parker, E. N. 1988, ApJ, 330, 474

The Era of Multi-Messenger Solar Physics
Proceedings IAU Symposium No. 372, 2023
G. Cauzzi & A. Tritschler, eds.
doi:10.1017/S1743921323000534

Study of the Impact of Coronal Rain on Kink Oscillations of Coronal Loops

Arpit Kumar Shrivastav[1,2,3] **and Vaibhav Pant**[1]

[1]Aryabhatta Research Institute of Observational Sciences, Nainital, 263001, India

[2]Joint Astronomy Programme, Department of Physics, Indian Institute of Science, Bangalore, 560012, India

[3]Indian Institute of Astrophysics, Bangalore, 560034, India

Abstract. Coronal rain occurs in thermally unstable coronal loops, and comprises cool plasma condensations, falling towards the solar surface, guided by the magnetic field. Sometimes the coronal rain clumps are seen to be subjected to transverse oscillations. The numerical simulations have indicated that coronal rain can onset kink oscillations in coronal loops and can affect the properties of oscillations. In this proceeding, we present the analysis of transverse oscillations in conjunction with coronal rain. Atmospheric Imaging Assembly (AIA) is used to examine the characteristics of coronal loop oscillation before and after coronal rain development. The analysis showed that the amplitude and period of oscillations are greater during coronal rain.

Keywords. magnetohydrodynamics (MHD) - Sun: corona - Sun: magnetic fields - Sun: flares - waves

1. Introduction

Transverse displacements of coronal loops are understood as standing kink modes (Edwin & Roberts 1983). They can be generated by nearby flare or any energetic event and can have decaying behaviour (White et al. 2012, Sarkar et al. 2016). The other type of kink oscillations, known as decay-less kink oscillations, can be persistent without any decaying behaviour and are very common in the solar corona (Anfinogentov et al. 2015). The responsible drivers for the generation of these oscillations are suggested as random footpoint motions or supergranular motions near footpoints (Nisticò et al. 2013, Nakariakov et al. 2016).

Other than the excitation mechanism of kink oscillations mentioned above, Verwichte & Kohutova (2017) presented the first observational evidence of excitation of kink oscillation by coronal rain production at the loop apex of a coronal loop. Their study revealed the variation of the oscillation properties due to the loop's evacuation caused by the descent of coronal rain clumps. Their study also indicated some signs of transverse oscillations before coronal rain in an overlying loop but not in the same loop due to the coarser resolution of AIA. The quantification of oscillation properties of a coronal loop before and after the development of coronal rain can provide information about the effect of coronal rain on kink oscillations.

In this article, we discuss the properties of kink oscillation before and after the production of coronal rain and examine the change in the properties of oscillations due to coronal rain.

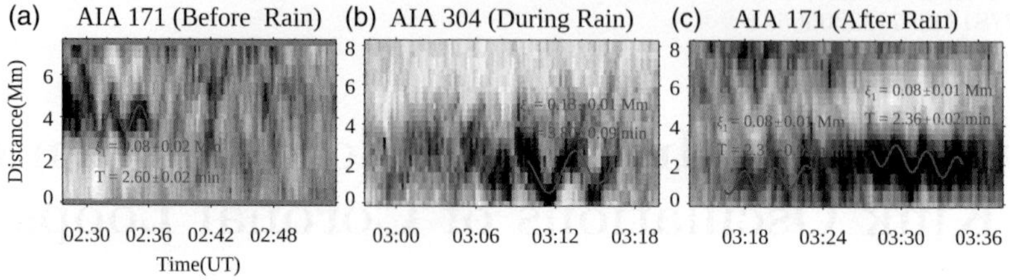

Figure 1. Motion Magnified x-t maps for Slit 3 in (a) AIA 171 before coronal rain, (b) AIA 304 during coronal rain, and (c) AIA 171 after coronal rain. The x-t maps are inverted in intensity. ξ_1 and T represent the amplitude and period of oscillations obtained after fitting.

2. Our Work

The event was observed in the 304 and 171 passbands of AIA and 1400 and 2796 passbands of Slit-Jaw Imager (SJI). These passbands allow us to track the oscillations at similar positions before, during, and after the formation of the coronal rain. The oscillations amplitudes detected in SJI passbands during coronal rain are below $\approx$ 1Mm, which is near the resolution of AIA. This fact motivated us to use the motion magnification technique (Anfinogentov & Nakariakov 2016) with a motion magnification factor of 7 to magnify the transverse oscillations in AIA passbands. We placed nine artificial slits in two coronal loops exhibiting transverse oscillations and coronal rain simultaneously. We obtained properties of these oscillations by fitting individual oscillating threads. Figure 1 shows an example of oscillations detected in Slit 3 before, during and after coronal rain.

3. Conclusions

AIA 171 x-t maps show that the oscillation periods and amplitudes are lower before and after coronal rain. The increase in amplitude during coronal rain can be due to its formation in a small time scale which can act as a sudden perturbation. The change in density profile along the loop due to the fall of coronal rain can be the reason behind the low period after coronal rain (Verwichte & Kohutova 2017, Kohutova & Verwichte 2017). For a detailed information on the work, please see this presentation.

Acknowledgements

A.K.S acknowledges the financial travel support that IAU granted to attend General Assembly XXXI of the International Astronomical Union in person. A.K.S is supported by the Council of Scientific & Industrial Research (CSIR), India, under file no. 09/079(2872)/2021-EMR-I. The authors also thank Ritesh Patel for valuable discussions.

References

Anfinogentov, S. & Nakariakov, V. M. 2016, Solar Physics, 291, 3251
Anfinogentov, S. A., Nakariakov, V. M., & Nisticò, G. 2015, A&A, 583, A136
Edwin, P. M. & Roberts, B. 1983, Solar Physics, 88, 179
Kohutova, P. & Verwichte, E. 2017, A&A, 606, A120
Nakariakov, V. M., Anfinogentov, S. A., Nisticò, G., & Lee, D. H. 2016, A&A, 591, L5
Nisticò, G., Nakariakov, V. M., & Verwichte, E. 2013, A&A, 552, A57
Sarkar, S., Pant, V., Srivastava, A. K., & Banerjee, D. 2016, Solar Physics, 291, 3269
Verwichte, E. & Kohutova, P. 2017, A&A, 601, L2
White, R. S., Verwichte, E., & Foullon, C. 2012, A&A, 545, A129

The Era of Multi-Messenger Solar Physics
Proceedings IAU Symposium No. 372, 2023
G. Cauzzi & A. Tritschler, eds.
doi:10.1017/S1743921322005099

Understanding Weak Impulsive Narrowband Quiet Sun Emissions (WINQSEs)

Divya Oberoi[1], Surajit Mondal[2], Rohit Sharma[3], Shabbir Bawaji[4], Ujjaini Alam[4] and Ayan Biswas[1,5,6]

[1]National Centre for Radio Astrophysics, Tata Institute of Fundamental Research, Pune, India

[2]New Jersey Institute of Technology, Newark, NJ, USA

[3]Fachhochschule Nordwestschweiz, Bahnhofstrasse 6, 5210 Windisch, Switzerland

[4]e4r^{TM}, ThoughtWorks, Pune, India

[5]Dept. of Physics, Engg. Physics & Astronomy, Queen's Univ., Kingston, Ontario, Canada

[6]Department of Physics, Royal Military College of Canada, Kingston, Ontario, Canada

Abstract. The confluence of data from the Murchison Widefield Array and an imaging pipeline tailored for spectroscopic snapshot images of the Sun at low radio frequencies have led to enormous improvements in the imaging quality of the Sun. These developments have lowered the detection thresholds by up to two orders of magnitude as compared to earlier studies, and have enabled the discovery of Weak Impulsive Narrowband Quiet Sun Emissions (WINQSEs). Their spatial distribution and various other properties are consistent with being the radio signatures of coronal nanoflares hypothesized by Parker (1988) to explain coronal heating in the quiet Sun emissions. We present the status of the multiple projects we have been pursuing to improve the detection and characterisation of WINQSEs, ranging from looking for them in multiple independent datasets using independent detection techniques to looking for their counter parts to estimate the energy associated with them and understanding their morphologies.

Keywords. Sun: corona, Sun: radio radiation

1. Introduction

The search for observational signatures of Parker's nanoflares, hypothesised to explain coronal heating, has usually focussed on detecting their thermal signatures in the EUV and X-ray bands. A convincing detection of nanoflares satisfying the requirements for coronal heating is yet to come by. The electron beams generated by these nanoflares are also expected to give rise to metrewave radio emission, though their detection had remained beyond the reach of most exist instrumentation. Recently, using data from the Murchison Widefield Array (MWA) (Tingay et al. 2013) and a highly optimised solar radio interferometric imaging pipeline (AIRCARS; Mondal et al. 2019), Mondal et al. (2020) presented the first detection of Weak Impulsive Narrowband Quiet Sun Emissions (WINQSEs). Their typical flux densities lie in the range of a few mSFU (1 SFU = 10,000 Jy) and they are found to occur in large numbers all over the Sun, including the quiet regions. In the solar radio images, they appear as compact sources and our estimate of their median duration is limited by the instrumental resolution of 0.5 s. These could potentially be the radio counterparts of these nanoflares. This work summarise the results and status from the follow up studies of WINQSEs.

2. Results, Conclusions and Future Directions

Given the implications of the discovery of WINQSEs and the challenges involved in detection of weak, transient, narrowband emissions, it is important to first put their detection on a firm footing. We have done so by establishing their presence in data taken during a period of exceptionally low solar activity alongside improving the imaging quality and detection methodology (Mondal et al. 2022, submitted). Apart from establishing their ubiquity and everpresence even during the very quiet solar data, it also helps better understand the distribution of WQINSEs. To further bolster the confidence in WINQSEs detection, we have used yet another dataset and an independent analysis procedure and software pipeline for their detection. Sharma et al. (2022) have detected WINQSEs in images made using the so called *residual visibilities* and their results are consistent with the other detections of WINQSEs. This work goes beyond earlier attempts as it also provides morphological information about the WINQSEs. From the coronal heating presepctive, the primary quantity of interest is the energy deposited in the corona. It is, however, very hard to derive this quantity from the radio observations, while this is routinely done using differential emission measure (DEM) techniques in the EUV band. While the EUV counterparts of individual WINQSEs seem to lie below the detection threshold of the current generation of instruments, Mondal (2021) identified the EUV counterpart of a group of WINQSEs. Using the DEM technique, he estimated the energy associated with it to be $\approx 3 \times 20^{25}$ergs, comfortingly close to the expectations from a group of nanoflares. As the numbers of WINQSEs detected runs into tens of thousand per hour per MHz, characterising their observational properties requires robust automated tools. For this we have developed and implemented a machine learning based approach for WINQSEs detection in solar radio images (Bawaji et al. 2022, submitted). In addition to significantly improving the detection methodology, this work also establishes that the bulk of isolated WINQSEs can be described well by a 2D Gaussian model and opens the interesting possibility of using the observed morphology of WINQSEs to constraint coronal turbulence and scattering at these otherwise hard to access coronal heights.

Ongoing and planned investigations of WINQSEs include estimating the bandwidth of some of the brighter WINQSEs, studying their time profile using higher time resolution data and their polarisation properties using the newly developed polarimetric imaging pipeline (Kansabanik et al. 2022). WINQSEs continue to meet all of the requirements for being the radio counterparts of Parker's nanoflares and we are carefully evaluating their role in coronal heating. We expect the the upcoming MWA Phase-III and the future Square Kilometer Array Observatory to enable the next big steps in this direction.

Acknowledgements

DO and AB acknowledges support of the Department of Atomic Energy, Government of India, under the project no. 12-R&D-TFR-5.02-0700. SM acknowledges partial support by USA NSF grant AGS-1654382 to the New Jersey Institute of Technology. RS acknowledges support of the Swiss National Foundation, under grant 200021 175832.

References

Kansabanik, D., Oberoi, D., & Mondal, S. 2022, ApJ, 932, 110
Mondal, S. 2021, Sol Phys, 296, 131
Mondal, S., Mohan, A., Oberoi, D., et al. 2019, ApJ, 875, 97
Mondal, S., Oberoi, D., & Mohan, A. 2020, ApJ Letters, 895, L39
Sharma, R., Oberoi, D., Battaglia, M., & Krucker, S. 2022, ApJ, 937, 99
Tingay, S. J., Goeke, R., Bowman, J. D., et al. 2013, Pub Ast Soc Aus, 30, e007

The Era of Multi-Messenger Solar Physics
Proceedings IAU Symposium No. 372, 2023
G. Cauzzi & A. Tritschler, eds.
doi:10.1017/S1743921323001199

Spatio-temporal Characterization of Hot Chromospheric Fibrils

Parker Lamb[1,2], **Gianna Cauzzi**[1,3] **and Kevin Reardon**[1,4]

[1]National Solar Observatory, Boulder, CO 80303, USA

[2]Department of Physics, University of Colorado, Boulder, CO 80309, USA

[3]INAF – Osservatorio Astrofisico di Arcetri, 50125 Firenze, Italy

[4]Department of Astrophysics and Planetary Sciences, University of Colorado, Boulder, CO 80303, USA

Abstract. The exact mechanisms leading to chromospheric heating are still ill-defined. While the presence of magnetic elements is undoubtedly necessary, the details of the heating, and its spatio-temporal distribution remain poorly understood. We contribute to this topic by analyzing the behavior of hot chromospheric fibrils surrounding network and plage elements, identified via the broader Hα profiles observed along their length; the H-α spectral line width has been shown to correlate with the local chromospheric temperatures through comparison with the ALMA millimeter-continuum brightness temperature. We make use of loop tracing and analysis software to investigate characteristics of the chromospheric hot fibrils including their length, number density, and transverse spatial extension in an enhanced network region.

Keywords. Sun: chromosphere, Sun: faculae, plage

1. Introduction

A clear explanation of the mechanisms leading to heating of the solar chromosphere still eludes us. The energy requirements are significant: on average, the chromosphere radiates 4 kWm^{-2} in quiet Sun, and 20 kWm^{-2} in active regions, as derived from the earlier work of Withbroe & Noyes (1977) and Vernazza et al. (1981). However, recent studies based on high resolution observations point to sensibly higher values, at least in some portions of plage and active regions (up to 90 and 160 kWm^{-2} respectively, see Morosin et al. 2022; Díaz Baso et al. 2021).

We know that the presence of magnetic elements is an important factor: beside active regions, both quiet Sun and plage regions show increased emission in areas where strong (photospheric) fields are present. The role of chromospheric magnetic fields has also recently become a subject of study, with Leenaarts et al. (2018) finding that enhanced radiative losses are correlated with the strength of the horizontal chromospheric fields. In general, however, analyses of the spatio-temporal distribution of chromospheric heating are still sparse, as they require high spatial and temporal resolution, long-duration observations as well as significant spectral and polarimetric information to derive the thermodynamical and magnetic properties of the atmosphere (see e.g. the recent study by Morosin et al. 2022).

A constant characteristics of areas of concentrated magnetic elements such as network or plage regions, is the presence of a multitude of fibrilar strutcures, anchored in the strong magnetic elements at the photosphere, as usually observed in the core of strong chromospheric spectral lines. These have been long observed for example using the

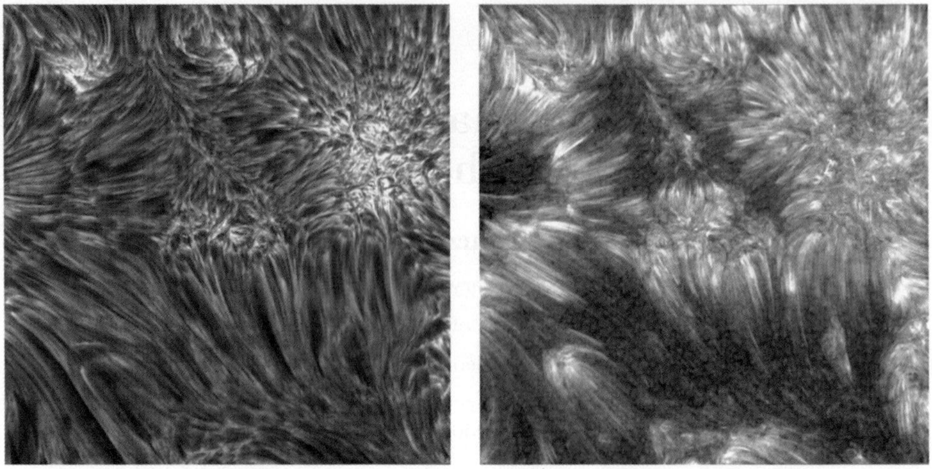

Figure 1. Left: Hα line core intensity in the analyzed plage area. The FOV is 96" × 96" (about 70 × 70 Mm). Right: Map of corresponding Hα spectral width. Values are scaled between 0.9 and 1.2 Å, which correspond roughly to T_b of 6-12,000 K, as per Molnar et al. (2019).

Hα line, and variously dubbed as "mottles", "rosettes", or spicules (Beckers 1968; Rutten 2006; Tsiropoula et al. 2012, among many). Their origin is still somewhat debated, but the important consideration for our present study is that such fibrils seem to be the site of strong heating events. This was originally recognized by Cauzzi et al. (2009), who showed how the immediate surroundings of network and plage elements would correspond to volumes of plasma hotter than the average, as indicated by the enhanced values of the spectral line widths of both Hα and the CaII 854.2 nm line. This finding was put on firmer footing by Molnar et al. (2019), that clearly demonstrated how regions of enhanced Hα spectral width coincide with area of high brightness temperature as inferred from the ALMA millimeter-continuum emission (T_b, which corresponds to the electronic temperature, T_e). The recent work of Tarr et al. (2023) further confirmed this correspondence.

Given the likely relevance of these "hot chromospheric fibrils" in the general problem of chromospheric heating, we have initiated a study of their quantitative characteristics, with the final objective of constraining possible heating mechanisms in this mysterious part of the solar atmosphere.

2. Analysis and results

The relevant data was acquired with the Interferometric Bidimensional Spectrometer (IBIS, Cavallini 2006) on April 23, 2017, and included sampling of the chromospheric lines Hα, NaI D_1 and CaII 854.2 nm in multiple wavelengths per line, with a cadence of ~16 s over a total of 46 minutes (note that this is the same dataset analyzed in Molnar et al. 2019; Hofmann et al. 2022). The 96" × 96" field of view (FOV) covered a stable plage area in the vicinity of an active region, as shown in the left panel of Figure 1, taken in the core of Hα. The right panel of Figure 1 shows instead the map of Hα spectral width, calculated from the whole spectral line as described in Cauzzi et al. (2009). A multitude of hot fibrils are clearly seen, jutting out of the plage magnetic elements. While the general orientation of the fibrilar features is the same in the two panels, note that hot fibrils (i.e. bright features in the map of Hα spectral width) can coincide with features that are either bright or dark in the intensity image. Moreover, the hot fibrils extend only for a certain fraction of the intensity fibrils, and do not cover the whole FOV.

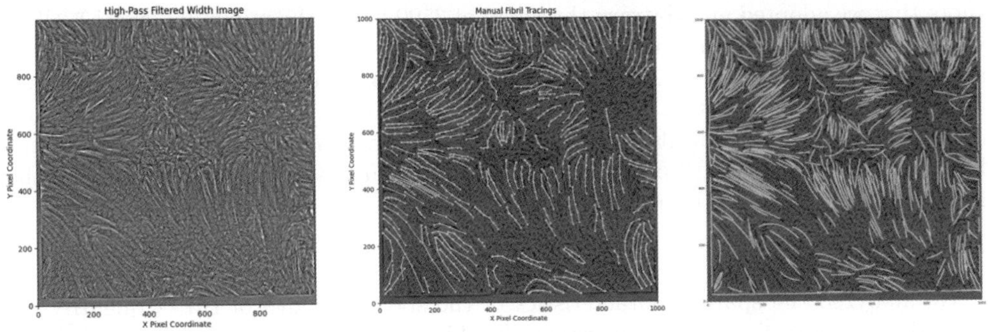

Figure 2. Left: High-pass difference image used to aid the manual identification of the hot fibrils. Center: Manual tracing (blue lines) of all visually-identifiable fibril lines in the width image (green background), for use as a control set. Right: Automatic tracing of fibrils by OCCULT-2 when using optimized parameters.

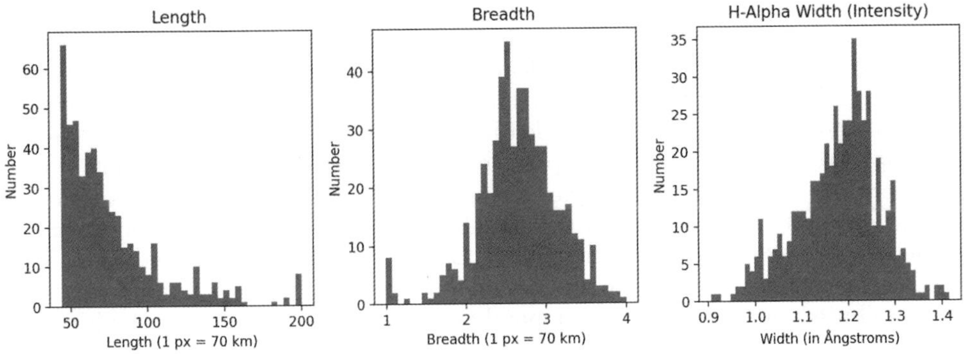

Figure 3. Left: Distribution of hot fibrils' lengths. A threshold of at least 50 pixels (3500 km) was used when selecting features. The average length of a hot fibril amounts to about 5500 km. Center: distribution of breadths. The average breadth of a hot fibril is of order 200 km, which is very close to the diffraction limit of the observations ($\approx$150 km). Right: distribution of Hα widths. Hot fibrils have an average Hα width of 1.18 Å, which following Molnar et al. (2019) would correspond to an approximate temperature of over 10,000 K.

The preliminary analysis described in this paper was conducted on a single map of Hα width, obtained at a time of excellent seeing. We first performed a manual identification of the hot fibrils, using a high-pass difference image obtained by subtracting a Gaussian-smoothed width map from the original image, for easier tracing of small-scale features (Fig. 2 left panel). The identification was performed multiple times, by multiple independent observers, to obtain a control set (Fig. 2 center panel). Finally, we used the OCCULT-2 algorithm (Aschwanden et al. 2013), which identifies curvilinear features from an intensity image and outputs a table of feature coordinates, to perform an automatic tracing of the hot fibrils. OCCULT-2 was first designed to trace coronal loops in images with lower spatial resolution, but we optimized its parameters using the control set defined above to guide our choices. The resulting traces are shown in Fig. 2 right panel. We identified a total of 515 fibrils in our 96" × 96" FOV.

Once we had the identified set of hot fibrils, we looked for three primary characteristics: **length** - the spatial extension of any given fibril; **breadth** - the thickness of a fibril at any given point along its length, perpendicular to the fibril's axis; **"intensity"** - the value of the Halpha spectral width at any given point along its length. The results are summarized in the histograms of Fig. 3. The hot fibrils display a large range of lengths (note that we adopted an arbitrary threshold of 3500 km as the minimum length), all the way to 10,000 km, with an average length of 5,500 km. Their average "intensity", e.g. Hα spectral width, is around 1.2 Å, which is close to the upper limit reported in Molnar et al. (2019), and corresponds to a T_e of over 10,000 K. Maybe most interesting, the breadth distribution shows that fibrils are very narrow, with values very close to the diffraction limit of the observations (2 pixels, or 150 km). This suggests that the mechanisms that lead to heating in these chromospheric features operate on very small spatial scales.

3. Conclusions

We analyzed the spatial characteristics of the spectral width of Hα, a quantity correlated with local T_e, in a plage region. To this end we have successfully adapted the OCCULT-2 algorithm, originally developed to trace coronal loops (Aschwanden et al. 2013), to identifiy curvilinear features at high spatial resolution. We find over 500 "hot fibrils" (i.e. features with a large Hα spectral width) in our 96" $\times$ 96" FOV, with a suggested temperatures in excess of 10,000 K, i.e. significantly hotter than the average chromosphere.

These features have an average length of around 5,500 km, and, while their general orientation is the same as for intensity features, their extension does not cover the whole FOV. This suggests that the heating events causing the large spectral widths occur only in a fraction of the fibrils as typically identified in the core of chromospheric lines, and, in particular, the heating occurs in the portion closer to their magnetic footpoints. Finally, their "breadth" (transversal extension) is of only 200 km, which indicates that these features are probably still under-resolved. Given these characteristics, and in particular the large aspect ratio of the hot fibrils, Ohmic dissipation of current sheets in the chromosphere appears a probable mechanism for the heating. However, a firmer interpretation will need an analysis of the temporal evolution of the heating, over multiple solar scenes.

Acknowledgements

P.L. was supported by the National Science Foundation REU program, award #1659878. G.C. acknowledges a IAU travel grant for participation in the General Assembly. The National Solar Observatory is operated by the Association of Universities for Research in Astronomy, Inc. (AURA), under cooperative agreement with the National Science Foundation. We are extremely grateful to the Local Organizing Committee for the organization of an excellent Assembly in spite of the many logistical difficulties.

Supplementary material

To view supplementary material for this article, please visit https://doi.org/10.1017/S1743921323001199.

References

Aschwanden, M., De Pontieu, B., & Katrukha, E. 2013, Entropy, 15, 3007
Beckers, J. M. 1968, Solar Physics, 3, 367
Cauzzi, G., Reardon, K., Rutten, R. J., Tritschler, A., & Uitenbroek, H. 2009, A&A, 503, 577
Cavallini, F. 2006, Solar Physics, 236, 415
Díaz Baso, C. J., de la Cruz Rodríguez, J., & Leenaarts, J. 2021, A&A, 647, A188
Hofmann, R. A., Reardon, K. P., Milic, I., et al. 2022, ApJ, 933, 244
Leenaarts, J., de la Cruz Rodríguez, J., Danilovic, S., Scharmer, G., & Carlsson, M. 2018, A&A, 612, A28
Molnar, M. E., Reardon, K. P., Chai, Y., et al. 2019, ApJ, 881, 99

Morosin, R., de la Cruz Rodríguez, J., Díaz Baso, C. J., & Leenaarts, J. 2022, A&A, 664, A8
Rutten, R. J. 2006, in Astronomical Society of the Pacific Conference Series, Vol. 354, Solar MHD Theory and Observations: A High Spatial Resolution Perspective, ed. J. Leibacher, R. F. Stein, & H. Uitenbroek, 276
Tarr, L. A., Kobelski, A. R., Jaeggli, S. A., et al. 2023, Frontiers in Astronomy and Space Sciences, 9, 978405
Tsiropoula, G., Tziotziou, K., Kontogiannis, I., et al. 2012, Space Science Reviews, 169, 181
Vernazza, J. E., Avrett, E. H., & Loeser, R. 1981, ApJS, 45, 635
Withbroe, G. L. & Noyes, R. W. 1977, ARA&A, 15, 363

SOC member Dipankar Banerjee from ARIES (India) asking a question during the Symposium.

Chapter IV:
MACHINE LEARNING IN SOLAR PHYSICS

Yong-Jae Moon from Kyung Hee University (Republic of Korea) delivering the review talk on current applications of machine learning in heliospheric studies.

The Era of Multi-Messenger Solar Physics
Proceedings IAU Symposium No. 372, 2023
G. Cauzzi & A. Tritschler, eds.
doi:10.1017/S1743921322004847

Application of Deep Learning to Solar and Space Weather Data

Yong-Jae Moon[1,2], Harim Lee[2], Jihyeon Son[1], Suk-Kyung Sung[2], Kangwoo Yi[2], Hyun-Jin Jeong[1], Eunsu Park[3], Eun-Young Ji[2], Il-Hyun Cho[2], Bendict Lawrance[2], Daye Lim[4], Gyungin Shin[5], Sujin Lee[6], Sumiaya Rahman[1] and Taeyoung Kim[1,7]

[1]School of Space Research, Kyung Hee University, Yongin, 17104, Republic of Korea

[2]Department of Astronomy and Space Science, College of Applied Science, Kyung Hee University, Yongin, 17104, Republic of Korea

[3]Korea Astronomy and Space Science Institute, Daejeon, 34055, Republic of Korea

[4]Centre for mathematical Plasma-Astrophysics, Department of Mathematics, KU Leuven Leuven, 3001, Belgium

[5]Department of Engineering Science, University of Oxford, Oxford, United Kingdom

[6] Space Center, Republic of Korea Air Force, Gyeryong, Republic of Korea

[7]AI Factory, Daejeon, Republic of Korea

Abstract. In this review, we introduce our recent applications of deep learning to solar and space weather data. We have successfully applied novel deep learning methods to the following applications: (1) generation of solar farside/backside magnetograms and global field extrapolation based on them, (2) generation of solar UV/EUV images from other UV/EUV images and magnetograms, (3) denoising solar magnetograms using supervised learning, (4) generation of UV/EUV images and magnetograms from Galileo sunspot drawings, (5) improvement of global IRI TEC maps using IGS TEC ones, (6) one-day forecasting of global TEC maps through image translation, (7) generation of high-resolution magnetograms from Ca II K images, (8) super-resolution of solar magnetograms, (9) flare classification by CNN and visual explanation by attribution methods, and (10) forecasting GOES solar X-ray profiles. We present major results and discuss them. We also present future plans for integrated space weather models based on deep learning.

Keywords. THE SUN, Sun: general, Sun: magnetic fields, methods: data analysis, techniques: image processing

---

1. Introduction

Deep learning (DL), which has been developed to find the best mathematical manipulation to turn the input into the output, is being applied to various fields and producing tremendous results. The technology has also been applied to astronomy and space science in recent years, helping research such as unprecedented data generation, damaged/lost data recovery, data quality improvement, event classification, detection, and development of space weather forecasting models.

A DL model for image-to-image translation generates different types of data from one or several input data. The Pix2Pix (Isola et al. 2017) model based on the conditional Generative Adversarial Networks (cGAN; Mirza & Osindero 2014) was suggested for the general purpose of image translation. In this review we apply this method for the generation of modern satellite data from distant past drawings, filling the data gap due to ground observation restrictions, removing noise, and converting a low-resolution image to a high-resolution image, generating the ionospheric Total Electron Content (TEC) map, 1-day forecasting of global TEC, etc. It is noted that there are several limitations such as artifacts found in high-resolution images by the Pix2Pix model. In this study, we have employed the image-to-image translation method (Pix2PixHD; Wang et al. 2018) for high-resolution images above 1 K. For scientific data applications, we develop a new image translation model, (Pix2PixCC; Jeong et al. 2022), with the inspector controlled by the correlation coefficient between the AI-generated image and the target image.

The Long Short-Term Memory (LSTM; Hochreiter & Schmidhuber 1997) is suitable for dealing with time-dependent data. Since the DL model, which has good performance in formalized data processing, was insufficient to solve problems with frequently changing I/O data, Sutskever et al. (2014) proposed a Sequence to Sequence (Seq2Seq) model based on LSTM using a general end-to-end approach to sequence learning with minimal assumptions about sequence structures. In this study, we predict the solar flare X-ray flux profile by designing the model to perform Seq2Seq frameworks using LSTM and attention mechanism.

Convolution Neural Networks (CNN), which are in the spotlight in image processing and computer vision, have already surpassed humans in their ability to classify and detect features many years ago. The excellent performance of CNN has been applied to develop space weather forecasting models such as the forecast of solar flare occurrence using full-disk solar magnetograms.

As artificial intelligence began to produce outstanding results, humans attempted to interpret the process of performing tasks inside the DL model. For CNN, a visual explanation such as guided backpropagation (Springenberg et al. 2014) and Gradient weighted Class Activation Mapping (Grad-CAM; Selvaraju et al. 2017) is possible. Here we develop a new DL flare forecasting model based on CNN to visually show where the important weights are distributed and compare them with the physical parameters of active regions.

In this review, we introduce various research results using novel DL methods conducted in the Sun and Space Weather Laboratory (SSWL) at Kyung Hee University (KHU). Section 2 introduces the studies using the image-to-image translation method, and Section 3 introduces space weather applications based on DL. A brief conclusion and prospects are given in section 4.

2. Image-to-Image Translation by Deep Learning

2.1. *Solar EUV images from magnetograms*

We apply novel DL methods to the image-to-image translation from solar magnetograms to solar ultraviolet (UV) and extreme UV (EUV) images (Park et al. 2019). For this, we consider two convolutional neural network models with different loss functions: one (Model A) is with L1 loss (L1), and the other (Model B) is with L1 and cGAN loss (LcGAN). We train the models using pairs of Solar Dynamics Observatory (SDO; Pesnell et al. 2012)/Atmospheric Imaging Assembly (AIA; Lemen et al. 2012) nine-passband (94, 131, 171, 193, 211, 304, 335, 1600, and 1700 Å) UV/EUV images and their corresponding SDO/Helioseismic and Magnetic Imager (HMI; Schou et al. 2012) magnetograms from 2011 to 2016. The models are evaluated by comparing pairs of SDO/AIA images and the corresponding ones generated in 2017. Our major results from this study are as follows. First, the models successfully generate solar UV and EUV images from SDO/HMI magnetograms (Figure 1). Second, the results from Model A are

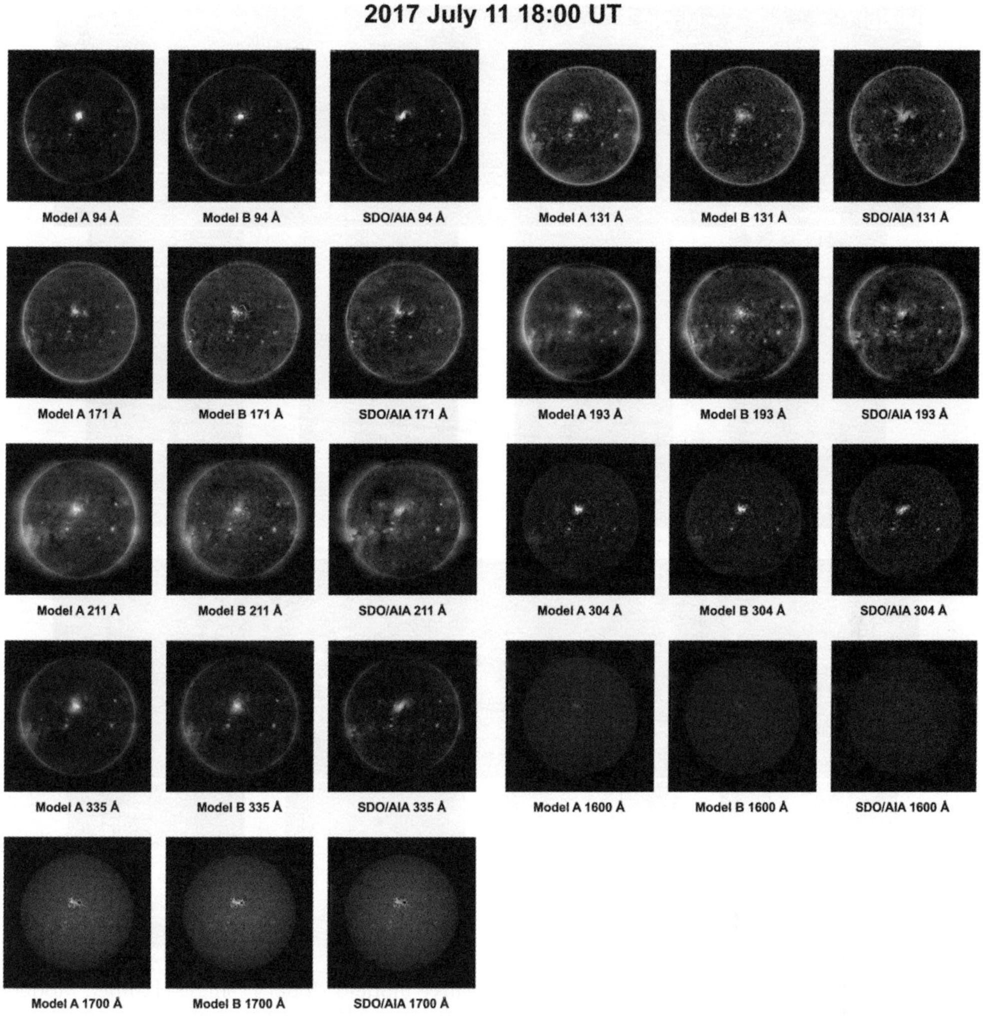

Figure 1. Comparison of real SDO/AIA images and the generated ones for nine passbands images at 12:00 UT on 2017 July 11. The first column shows generated images by Model A, the second column, the generated images by Model B, and the third column, SDO/AIA images (adapted from Park et al. (2018)).

slightly better than those from Model B in view of three metrics (pixel-to-pixel correlation coefficient (CC), relative error, and the percentage of pixels having errors less than 10 %). Third, in view of the RMS contrast measure, the generated images by Model A are much more blurred, which is due to the characteristics of convolution, than those by Model B because LcGAN is specialized for generating realistic images.

2.2. *Solar UV/EUV images from other UV/EUV images*

We use the above successful method on SDO/AIA 9 UV/EUV channels to find out which channel (or combination of channels) generated the most realistic image (Lim et al. 2021). 170 DL models with single-channel output (72 single-channel input models, 56 dual-channel input models, and 42 triple-channel input models) were developed and we evaluated the models with the pixel-to-pixel CC for the solar disk. Among the single-channel input models, the model with 131Å, among the dual-channel input models, the

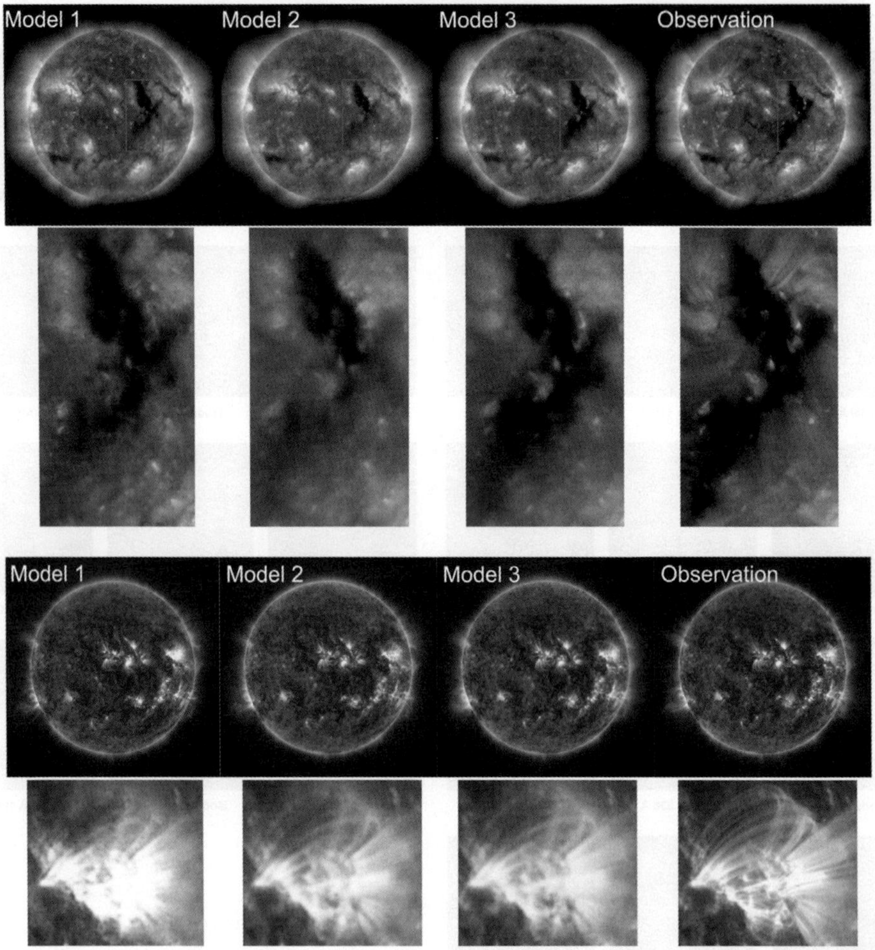

Figure 2. Comparison images between real and three model (Model 1: single-channel; Model 2: dual-channel; Model 3: triple-channel) results for CH and AR. The first and third rows are full disk images of 193Å and 171Å with CH and AR. The second and fourth rows show the regions (red boxes) magnified in the row above. The pixel-to-pixel CCs obtained from the three models are 0.92, 0.92, and 0.96 for CH, and 0.89, 0.96, and 0.97 for AR, respectively (adapted from Lim et al. (2021)).

model with 131Å and 1600Å, and among the triple channel input models, the model with 131Å, 1600Å and 304Å have the highest average CC of 0.84, 0.95, and 0.97, respectively. The three passbands of the triple-channel model with the best performance are the channels that observe the coronal, upper photospheric, and chromospheric regions, respectively (Figure 2). This result may be taken into account when selecting channels for UV/EUV imaging instruments on solar satellites.

2.3. *Farside solar magnetograms from EUV images*

Magnetograms are significant for studying solar activity and predicting space weather disturbances. Farside/backside magnetograms can be constructed from local helioseismology without any farside data, but their quality is quite lower than that of typical frontside magnetograms. We generate farside solar magnetograms from Solar Terrestrial Relations Observatory (STEREO; Kaiser et al. 2008)/EUV 304 Å images using a DL

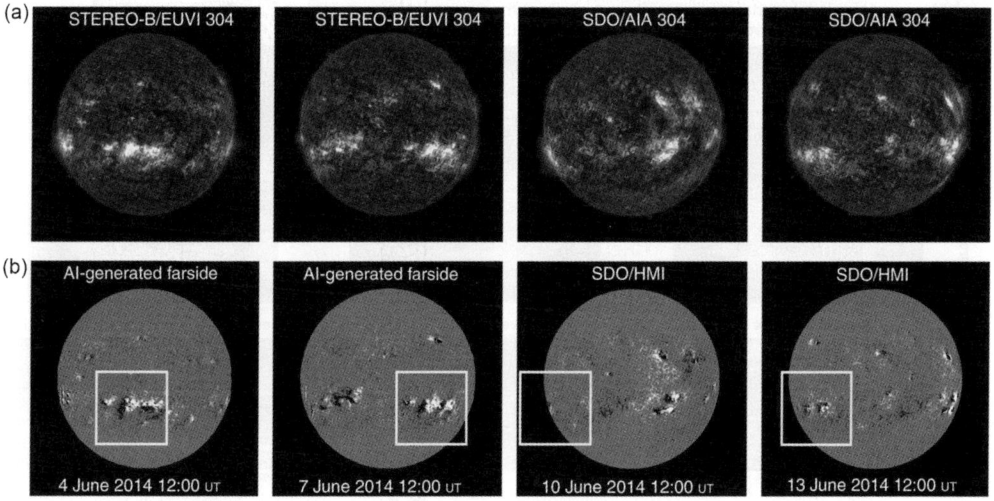

Figure 3. A series of 304Å images and magnetograms: a) the first two 304Å images are taken from STEREO-B/EUVI and the last two from SDO/AIA; b) the first two magnetograms are AI-generated farside ones from the model and the last two are SDO/HMI. The yellow boxes indicates the tracking of the NOAA active region 12087 from the farside to the frontside. We use ±100 G as the saturation level for both SDO/HMI and AI-generated magnetograms (adapted from Kim et al. (2019)).

model based on cGANs (Kim et al. 2019; hereafter KPL19). We train the model (Artificial Intelligence Solar Farside Magnetogram; AISFM v1.0) using pairs of SDO/AIA 304 Å images and SDO/HMI magnetograms taken from 2011 to 2017 except for September and October each year. The model is evaluated by comparing both SDO/HMI magnetograms and cGAN-generated magnetograms in September and October. Our method successfully generates frontside/backside solar magnetograms from SDO/AIA 304 Å images, and these are similar to those of the SDO/HMI. Thus we are able to monitor the temporal evolution of solar active regions from the farside to the frontside of the Sun using SDO/HMI (Figure 3) and farside magnetograms generated by our model when the farside EUV data are available.

Global magnetic fields of the solar corona are typically extrapolated from photospheric field data, for which farside data are taken when the AR was at the frontside, about two weeks earlier. For the first time, we construct the extrapolations of global magnetic fields using frontside and AI-generated farside magnetograms on a near-real time basis (Jeong et al. 2020). We generate the farside magnetograms from three channel images of STEREO Ahead (A) and Behind (B) by our DL model (AISFM v2.0) trained with frontside SDO EUV images and magnetograms. For frontside testing data sets, we demonstrate that the generated magnetic field distributions are quite consistent with the real ones; ARs as well as quiet regions (Figure 4). We make global magnetic field synchronic maps in which conventional farside data, which were taken many days ago, are replaced by farside ones generated by our model. The synchronic maps well produce not only the appearance of ARs but also the disappearance of others on the solar surface, which are much better than the conventional synoptic maps. We use these synchronized magnetic data to extrapolate the global coronal fields using the Potential Field Source Surface (PFSS) model. Our results are much more consistent with coronal EUV observations than those of the conventional method in view of solar active regions and coronal holes (Figure 5).

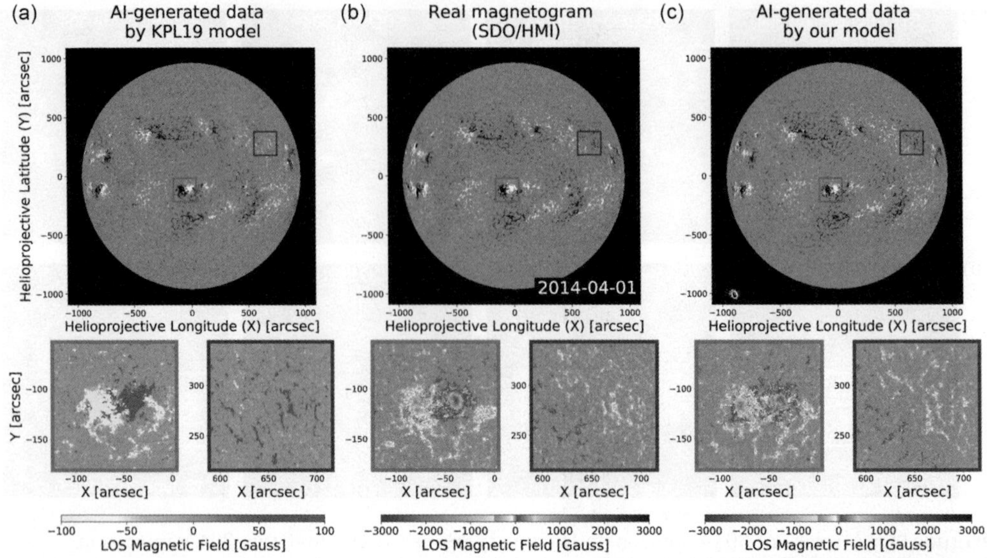

Figure 4. Comparisons of a real SDO/HMI magnetogram and AI-generated one: (a) AI-generated data by KPL19 model; (b) Real magnetogram on 2014 April 1 at 12:00 UT; (c) AI-generated one by the model. Full-disk magnetograms are displayed as white for positive polarity and black for negative one. A solar AR at disk center (red box) and a QR on the limb (blue box) are zoomed and represented with other color maps showing the large dynamic range values in Gauss (adapted from Jeong et al. (2020)).

We greatly improve AI-generated solar farside magnetograms using data sets from the STEREO and SDO (Jeong et al. 2020). We upgrade our previous DL model and input data sets to generate more realistic magnetograms than before. First, our model (AISFM v3.0), which is called Pix2PixCC (Figure 6), uses updated objective functions, which include CCs between the real and generated data. Second, we construct input data sets of our model: solar farside STEREO EUV images together with nearest frontside SDO data pairs of EUV ones and magnetograms. We expect that the frontside image pairs provide historic information on the distribution of magnetic field polarities. We demonstrate that magnetic field distributions generated by our model are much more consistent with the real ones than before, in view of several metrics. The averaged pixel-to-pixel CC for full disk, active regions, and quiet regions between real and AI-generated magnetograms with 8×8 binning are 0.88, 0.91, and 0.70, respectively. Total unsigned magnetic fluxes and net magnetic fluxes of the AI-generated magnetograms are consistent with those of real ones for the test data sets. It is interesting to note that our farside/backside magnetograms produce polar field strengths and magnetic field polarities consistent with those of nearby frontside magnetograms for solar cycles 24 and 25. Now we can monitor the temporal evolution of active regions using solar farside/backside magnetograms by the model together with the frontside ones. Our AI-generated solar farside magnetograms are now publicly available at the Korean Data Center for SDO (`http://sdo.kasi.re.kr`).

2.4. *Solar magnetograms and EUV images from sunspot drawings*

We generate solar magnetograms and EUV images from Galileo sunspot drawings using a DL model based on cGANs (Lee et al. 2021). We train the model using pairs of sunspot drawings from the Mount Wilson Observatory and their corresponding magnetograms (or EUV images) from 2011 to 2015 except for every June and December by

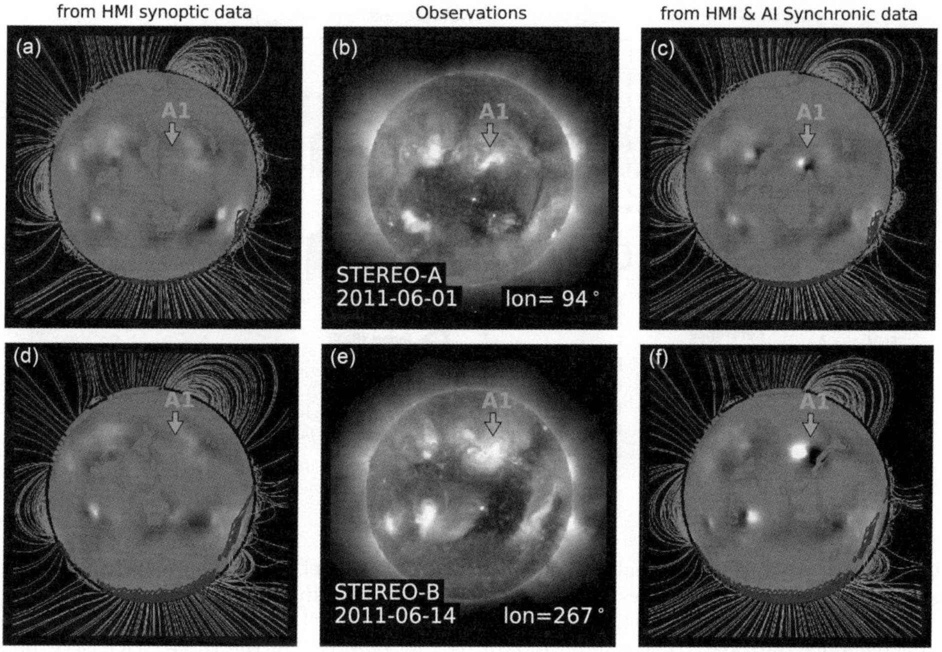

Figure 5. Comparisons of EUV observations and PFSS extrapolations from conventional and HMI and AI synchronic data (by our model) in full-disk observations. (a)-(c) Result of PFSS extrapolation from HMI synoptic data, EUV 284Å images, and PFSS extrapolations from SDO/HMI and AI synchronic data at the position of STEREO A on 2011 June 1. (d)-(f) Those at the position of STEREO-B on 2011 June 14. Positive and negative polarities of the open fields are displayed with blue and red colors. Closed field lines are indicated with dark yellow. Field lines at the limb and open field area on the surface are only shown for comparison. The PFSS is computed with source surface at $2.0R_\odot$ and $2.5R_\odot$, and those open field areas are displayed with hatched pattern and filled area, respectively. The solar surface is filled with the bottom boundary data to show the distribution of ARs. The green arrow indicates the appearance of the NOAA AR 11236 (adapted from Jeong et al. (2022)).

the Solar Dynamic Observatory satellite. We evaluate the model by comparing pairs of actual magnetograms (or EUV images) and the corresponding AI-generated image pairs in June and December. Our results show that bipolar structures of the AI-generated magnetograms are consistent with the original ones and their unsigned magnetic fluxes (or intensities) are consistent with the original ones. Applying this model to the Galileo sunspot drawings in 1612 (Galilei et al. 1613), we generate HMI-like magnetograms and AIA-like EUV images of the sunspots (Figure 7). We hope that the EUV intensities can be used for estimating solar EUV irradiance at long historical times. This study was publically introduced by AAS NOVA and Astronomy magazine.

2.5. *Solar magnetograms from Ca II images*

We generate realistic high-resolution (1024×1024 pixels) pseudo-magnetograms from Ca II K images using a DL model based on conditional generative adversarial networks (Shin et al. 2020). We consider a model "Pix2PixHD" that is specifically devised for high-resolution image translation tasks. We use Ca II K 393.3 nm images from the Precision Solar Photometric Telescope at the Rome Observatory (Ermolli et al. 1998, 2007) and magnetograms from the HMI at the SDO from 2011 January to 2015 June. 2465 pairs of Ca II K and HMI are used for training except for January and July data. The remaining

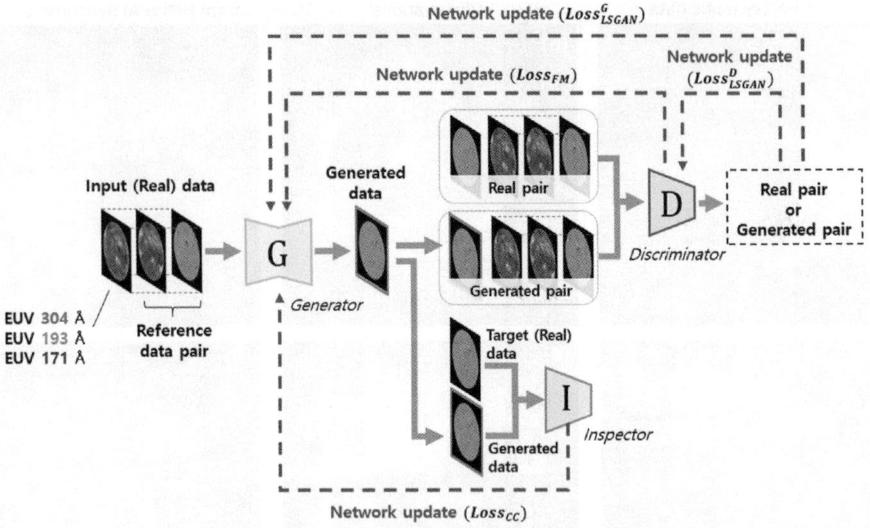

Figure 6. Flowchart and structures of the Pix2PixCC model. G, D, and I are the generator, discriminator, and inspector, respectively. The generator produces target-like data from input data. When we train the model, input data are SDO/AIA EUV 304, 193, and 171 Å images and a reference data pair. The reference data pair is composed of the three SDO EUV images and the corresponding SDO/HMI magnetogram, which are taken at the nearest available time. The discriminator trains for distinguishing between the real pair and generated ones. The real pair is composed of input data and target data, and the generated pair consists of input data and generated data. The inspector computes CCs between the target data and generated data. The generator and discriminator are updated by the losses calculated from the inspector and discriminator (adapted from Jeong et al. (2022)).

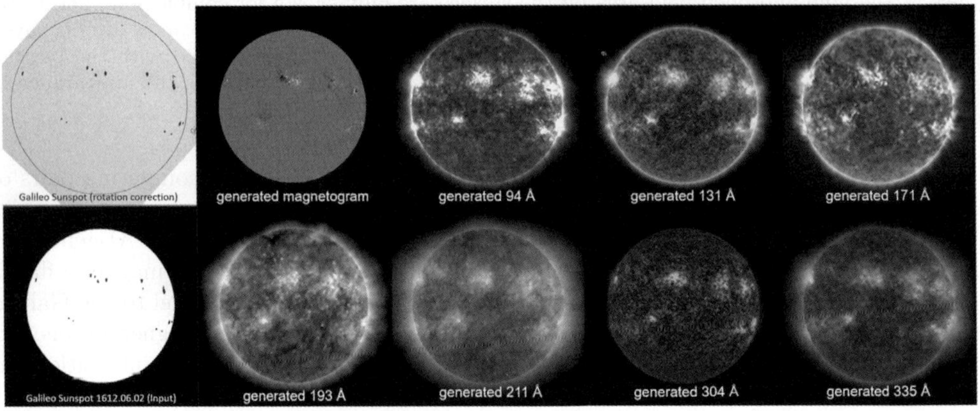

Figure 7. Results of our model when the Galileo sunspot drawings are used for input data: AI-generated magnetogram and UV/EUV images from the Galileo sunspot drawing on 1612 June 2 (modified from Lee et al. (2021a)).

436 pairs are used for an evaluation of the model. Our model shows that the average CC of total unsigned magnetic flux between AI-generated and real ones is 0.99 and the mean pixel-to-pixel CC after 8×8 binning over the full disk is 0.74. We find that the AI-generated absolute magnetic flux densities are very consistent with real ones, even in the fine-scale structures of quiet regions (Figure 8). On the other hand, the average pixel-to-pixel correlations of magnetic flux densities strongly depend on a region of interest: 0.81

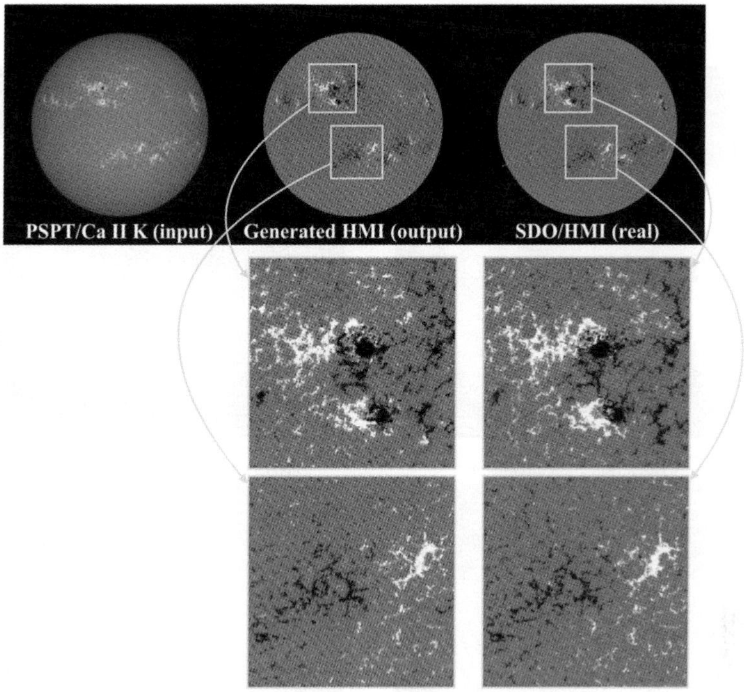

Figure 8. (Upper left) a Ca II K input intensity image, (upper middle) a generated HMI-like magnetogram, and (upper right) the corresponding real SDO/HMI. Ca II K and real SDO/HMI are taken at 18:48 UT 2012 January 31. (Lower left column) enlarged views of active regions in the generated magnetogram. (Lower right column) those in the real magnetogram. The white area represents positive polarities and the black area represents negative ones. For visual purpose, magnetic flux densities are expressed within ±100 G (adapted from Shin et al. (2020)).

for active regions and 0.24 for quiet regions. Our results suggest a sufficient possibility that we can produce high-resolution solar magnetograms from historical Ca II data.

2.6. *Solar He I 1083 nm images from EUV images*

We generate He I 1083 nm images from SDO/AIA images using a Pix2PixHD model based on cGAN (Son et al. 2021). He I 1083 nm images, the target data, were obtained from National Solar Observatory (NSO)/Synoptic Optical Long-term Investigations of the Sun (SOLIS; Keller et al. 2003). We make three DL models: single-input SDO/AIA 19.3 nm image for Model I, single-input 30.4 nm image for Model II, and double-input (19.3 and 30.4 nm) images for Model III. We use the data from 2010 October to 2015 July except for June and December for training and the remaining one for test. Major results of our study are as follows. First, the models successfully generate He I 1083 nm images with good correlations. Second, Model III has better results than those with single input image in view of metrics such as root mean square error (RMSE) and correlation coefficient (CC). RMSE and CC between real and synthetic ones for model III after 4 by 4 binnings are 9.49 and 0.88, respectively. Third, the AI-generated images show well observational structures such as active regions, filaments, and coronal holes. By our model, He I 1083 nm images can be generated with high time resolutions without data gaps (Figure 9), which would be useful for studying the time evolution of solar activity shown in He I 1083 nm observations.

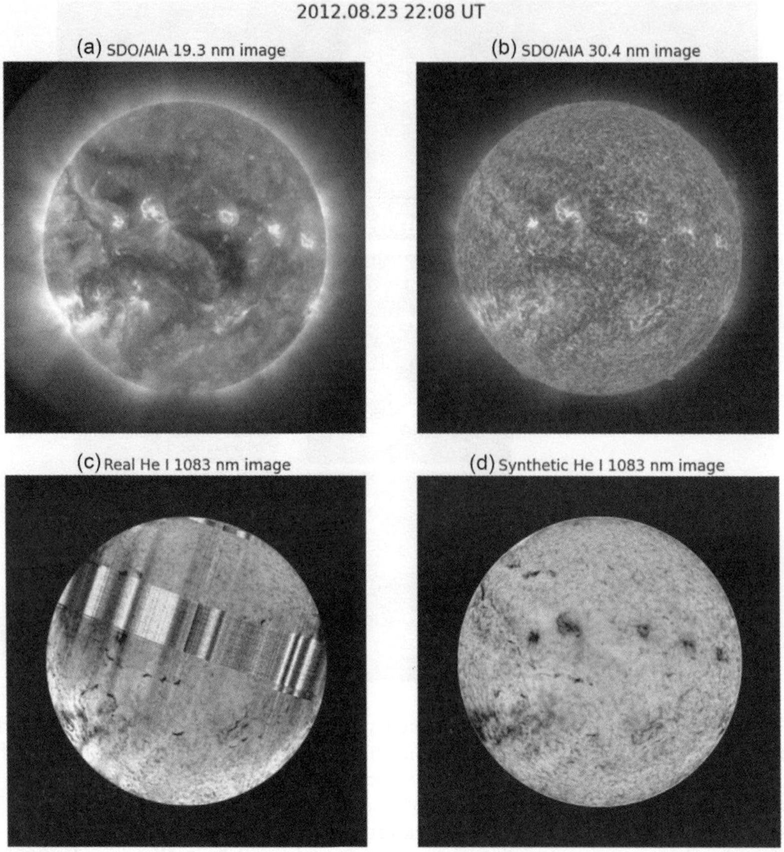

Figure 9. Restoration of damaged He I 1083 nm data at 22:08 UT on 2012 August 23 by our model: (a) SDO/AIA 19.3 nm image as input; (b) SDO/AIA 30.4 nm image as input; (c) observed He I 1083 nm image; (d) synthetic image from Model III (adapted from Son et al. (2021)).

2.7. *Solar coronal white-light images from EUV images*

The off-limb coronal white light images with an altitude range between 1.11 – 1.25 solar radii were generated from a single EUV channel image as well as multiple channel images by applying the DL model based on the cGANs (Lawrance et al. 2022). It was found that the AI-generated white light image shows the best correlation with the original K-Cor (de Wijn et al. 2012) image when multiple channels of the AIA 193 Å and 211 Å were used as input. The majority of low coronal structures such as helmet streamers, pseudostreamers, and polar coronal holes are well identified in the AI-generated ones as in the original images. It was also found that the positions and sizes of the polar coronal holes in the AI-generated images are well consistent with those of the original ones (Figure 10). In addition, the coronal mass ejections and jets were also well identified in the AI-generated images. Hence, the translation model from EUV images to white light images using the cGANs can provide complementary data to study the off-limb low coronal features in the white light when no ground-based observation is available due to the repeated night time, poor atmospheric conditions, and instrumental maintenance.

2.8. *Denoising solar magnetograms*

In astronomy, long-exposure observations are one of the important ways to improve signal-to-noise ratios (S/Ns). We apply a deep-learning model to de-noise solar

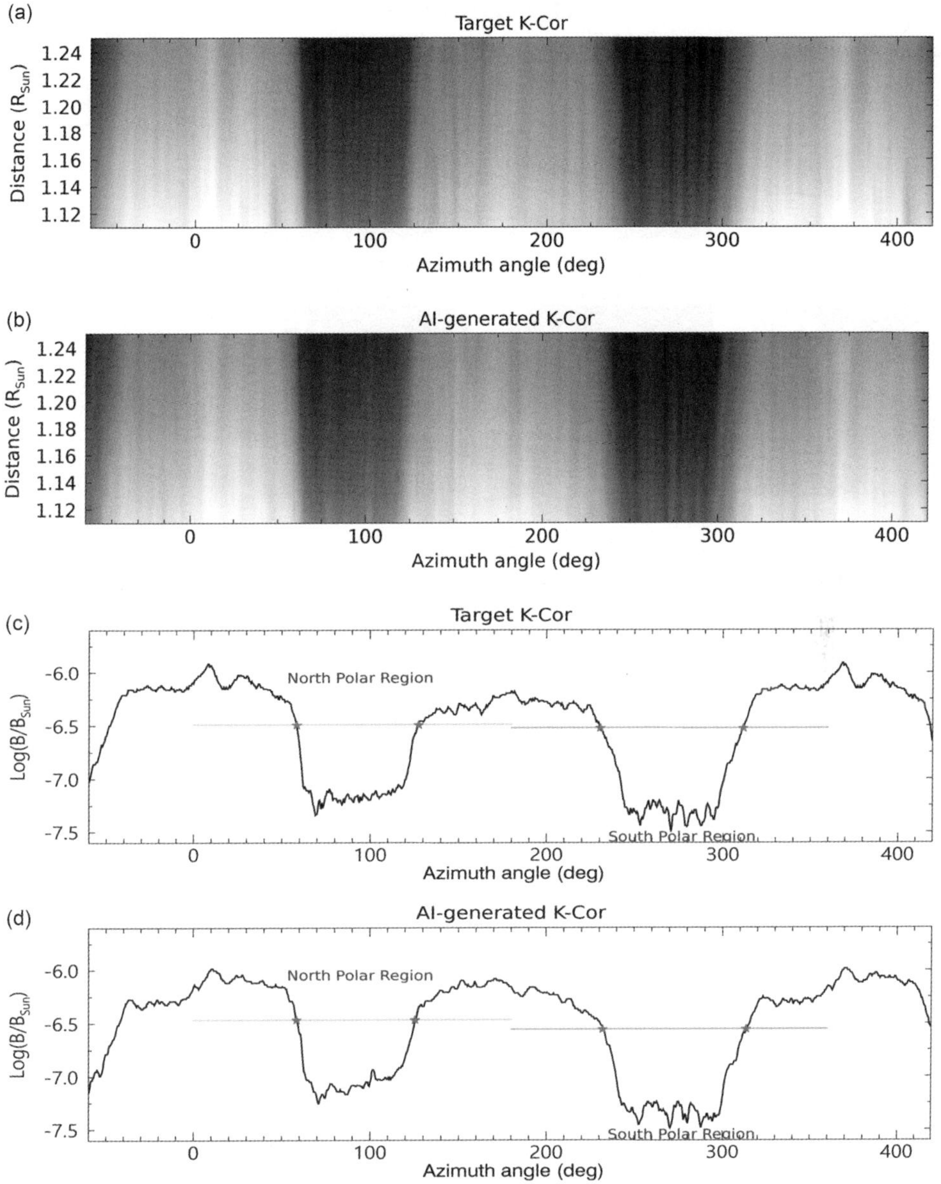

Figure 10. The white light images of the off-limb solar corona observed by K-Cor at 2018-Nov-27 19:38:52 UT (a) and its AI-generated image (b), corresponding intensity profiles averaged in the altitude range from 1.11 to 1.25 solar radii along azimuth angle (c, d). Green and orange lines are mean intensities of the northern and southern off-limbs, respectively (adapted from Lawrance et al. (2022)).

magnetograms (Park et al. 2020). This model is based on a deep convolutional generative adversarial network with a conditional loss for image-to-image translation from a single magnetogram (input) to a stacked magnetogram (target). For the input magnetogram, we use SDO/HMI line-of-sight magnetograms at solar disk center. For the target magnetogram, we make 21-frame-stacked magnetograms, taking into account solar rotation. We train the model using 7004 pairs of the input and target magnetograms from 2013 January to 2013 October. We then validate the model using 707 pairs from 2013 November and test the model using 736 pairs from 2013 December. Our major results

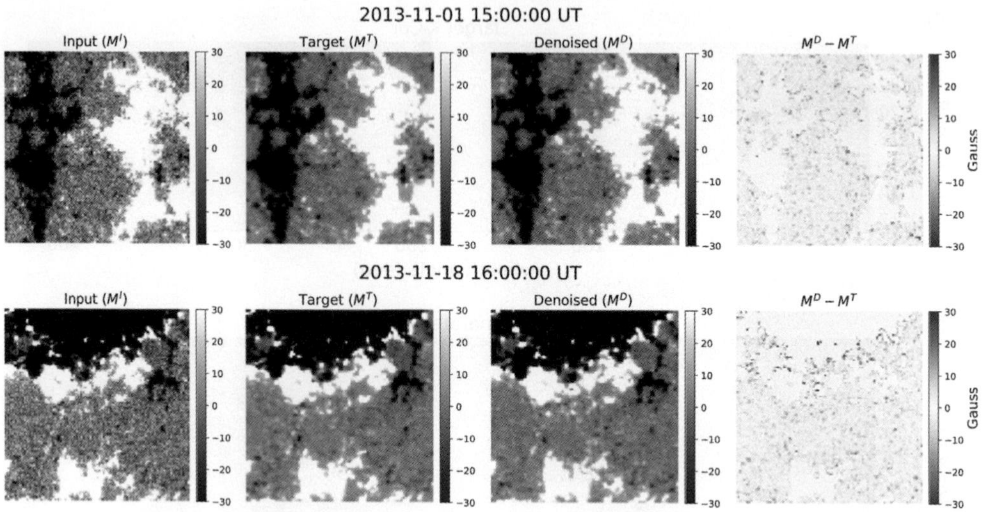

Figure 11. Two examples of comparisons of input, target, and de-noised magnetograms. The first column represents input magnetograms, the second column, target magnetograms, the third column, de-noised magnetograms from our model, and the last column, the difference maps between de-noised and target ones (adapted from Park et al. (2020)).

from this study are as follows. First, our model successfully denoises SDO/HMI magnetograms, and the de-noised magnetograms from our model are well consistent with the target ones (Figure 11). Second, the average noise level of the de-noised magnetograms is greatly reduced from 8.66 to 3.21 G, and it is consistent with that of the target magnetograms, 3.21 G. Third, the average pixel-to-pixel correlation coefficient value significantly increases from 0.88 (input) to 0.94 (de-noised), which means that the de-noised magnetograms are more consistent with the target ones than the input ones. Our results would be applied to many scientific fields in which the integration of many frames (or long-exposure observations) are used to increase the S/N.

2.9. *Improvement of IRI TEC maps*

We develop a DeepIRI model that generates by improving the total electron content (TEC) maps of the International Reference Ionosphere (IRI; Bilitza et al. 2017) model using DL based on conditional Generative Adversarial Networks (Ji et al. 2020). We use pairs of IRI TEC maps and International Global Navigation Satellite Systems (GNSS) Service (IGS; Kumar et al. 2015) TEC maps from 2001 to 2011 for training the model and compare IGS TEC and DeepIRI TEC maps from 2013 to 2017 for evaluating the model. The DeepIRI TEC maps generated from our model are more similar to the IGS TEC maps than IRI TEC maps, and in particular, ionospheric peak structures generate much more successfully in the DeepIRI TEC model than the IRI-2016 model (Figure 12). Comparing the average differences with the IGS TEC maps, the DeepIRI model significantly improves the IRI TEC at low-latitude regions around the equatorial anomaly. These results demonstrate the potential of the DeepIRI model based on deep learning to improve the global TEC prediction ability of the IRI model.

2.10. *Super-resolution of solar magnetograms*

Image super-resolution is a technique of enhancing the resolution of an image where a high-resolution (HR) image is reconstructed from a low-resolution (LR) image. We

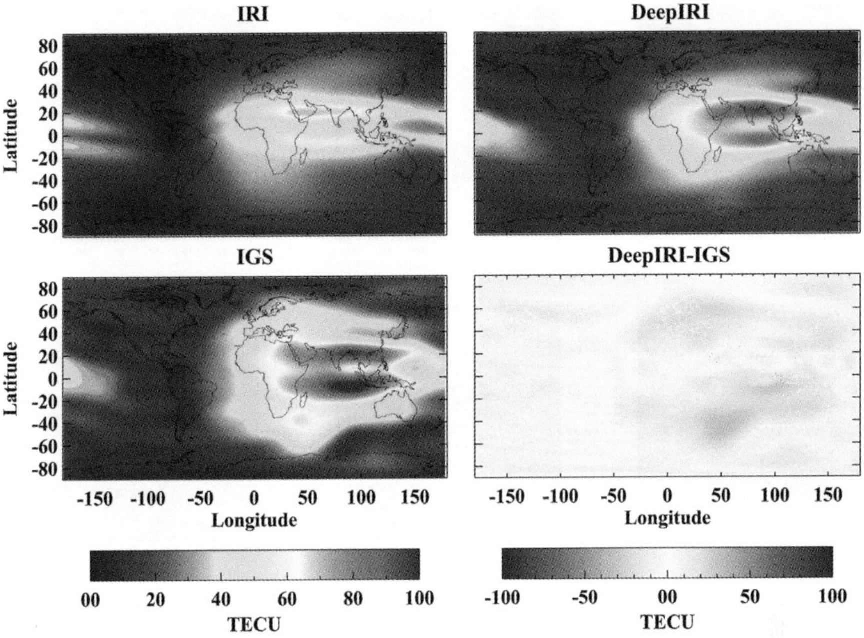

Figure 12. Example of the IRI TEC map, DeepIRI TEC map, IGS TEC map, and the difference map between the DeepIRI TEC and IGS TEC maps at 10 UT on 17 March 2015 (adapted from Ji et al. (2020)).

apply two novel DL models (residual attention model and progressive GAN model) to SDO/HMI magnetograms for super-resolution (Rahman et al. 2020). For this, we consider SDO/HMI0 magnetograms as output and their degraded ones with 4×4 binning as input. DL networks try to find internal relationships between LR and HR images from the given input and the corresponding output image. We use SDO/HMI magnetograms from 2014 May to August for training, from 2014 October to December for validation, and 2015 January to March for testing. We find that the DL models generate higher-quality results than the bicubic interpolation in terms of visual inspection and metrics (Figure 13). We apply this model to a full-resolution SDO/HMI magnetogram and then compare the generated magnetogram with the corresponding Hinode/SOT Narrowband Filtergrams (NFI) magnetogram. This comparison shows that the generated magnetogram is quite consistent with the Hinode one with a very good correlation (CC: 0.94) and a high similarity (SSIM: 0.93), which are better than the bicubic method.

3. Space Weather Forecast Based on Deep Learning

3.1. *Forecast of solar flare occurrence*

We apply CNN to the forecast of solar flare occurrence (Park et al. 2018). For this, we consider three CNN models (two pre-trained models, AlexNet and GoogLeNet, and one newly proposed model). Our inputs are the Solar and Heliospheric Observatory (SOHO; Domingo et al. 1995)/Michelson Doppler Imager (MDI; Scherrer et al. 1995) (from 1996 May to 2010 December) and SDO/HMI (from 2011 January to 2017 June) full-disk magnetograms at 00:00 UT. Model outputs are "Yes or No" of daily flare occurrence (C, M, and X classes) and they are compared with GOES flare identifications. We train the models using the input data and observations from 1996 to 2008, covering the entire solar cycle 23, and test them using the data sets from 2009 to 2017, covering solar cycle 24. Then we compare the results of our CNN models with those of three previous flare

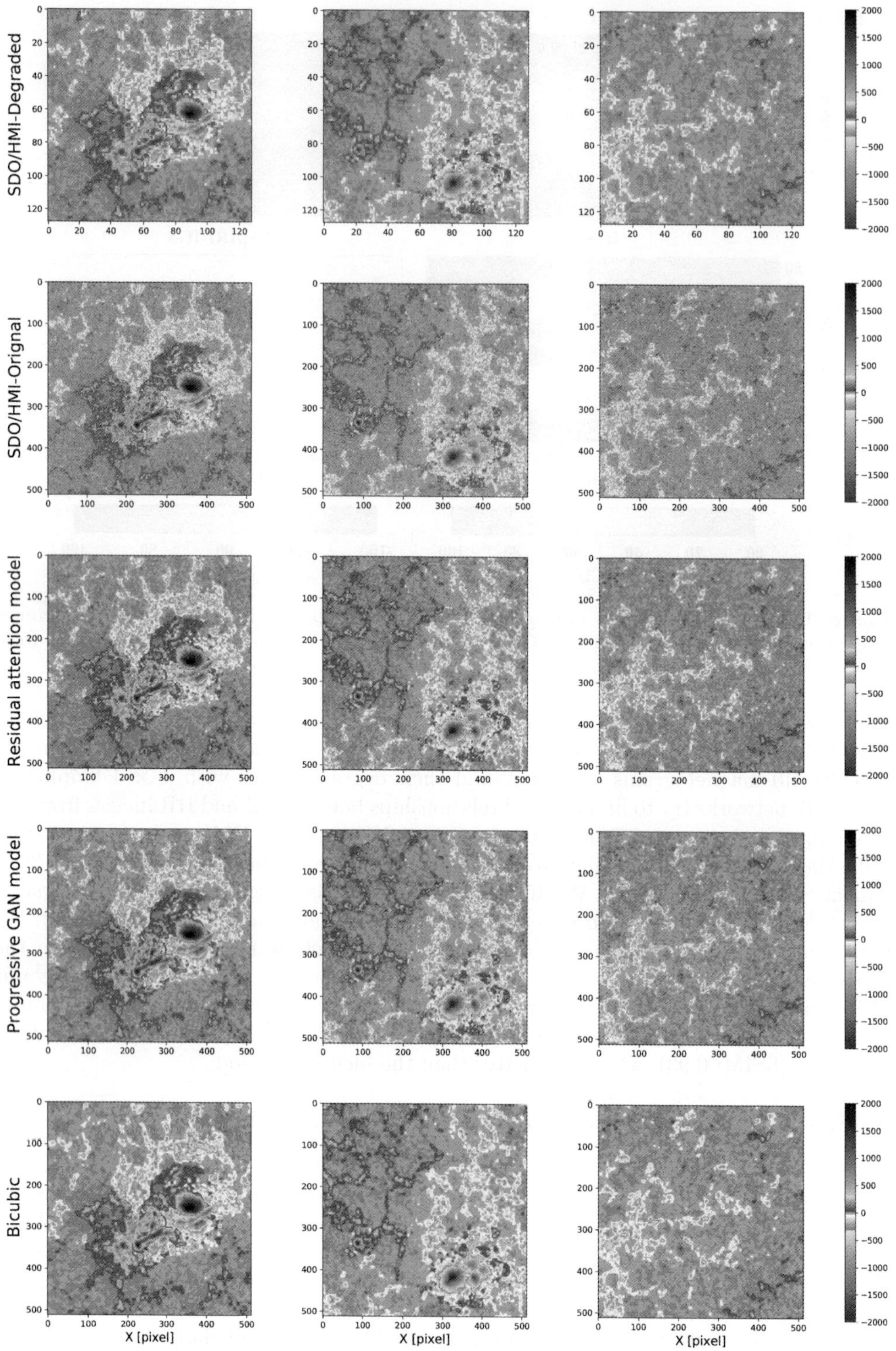

Figure 13. Results of DL models for super-resolution: input (first row), target (second row), AI-generated magnetograms (third and fourth row), and bicubic interpolation results (fifth row) for strong (first and second columns) and weak (third column) field regions. The target and input images are the original SDO/HMI magnetogram and its degraded low-resolution one, respectively (adapted from Rahman et al. (2020)).

Table 1. The Results of our CNN Models and their comparison with three previous flare forecast models (adapted from Park et al. (2018)).

	This work			Colak & Qahwaji (2009)	Bloomfield et al. (2012)	Huang et al. (2018)
	Model 1	**Model 2**	**Model 3**			
ACC	0.78	0.79	0.82	0.81	0.71	0.76
POD	0.72	0.84	0.85	0.81	0.75	0.73
CSI	0.64	0.84	0.85	...	...	0.31
FAR	0.14	0.21	0.17	0.30	0.65	0.65
HSS	0.57	0.57	0.63	0.51	0.32	0.34
TSS	0.57	0.56	0.63	...	0.46	0.49
Data sets	Training: 1996-2008 Test: 2009-2017			Training: 1972-1998 Test: 1999-2002	Training: 1988-1996 Test: 1996-2010	Training: 1996-2010 Test: 2010-2015

forecast models in view of statistical scores. The major results from this study are as follows. First, we successfully apply CNN to the full-disk solar magnetograms without any preprocessing or feature extraction. Second, the results of our CNN models are considerably better in false alarm ratio (FAR) and critical success index than those of the previous solar flare forecasting models (Table 1). Third, our proposed model has better values of all statistical scores except for FAR, than the other two pre-trained models. Our results show a sufficient possibility that DL methods can improve the capability of not only the solar flare forecast but also similar types of forecast problems.

3.2. *Forecast of major solar X-ray flare flux profiles*

We make a DL model to forecast major solar flare X-ray flux profiles, based on a Seq2seq and an attention mechanism (Yi et al. 2020). The model forecasts 30 minutes of X-ray flux profiles. We train the model using 845 solar flare X-ray data from 1998 to 2006. The model performance is evaluated with RMSE. We compare the model with six other models based on deep learning methods and regression methods. The major results of this study are as follows. First, the model successfully forecasts solar flare X-ray flux profiles (Figure 13), without any preprocessing to extract features from data. Second, our proposed model achieves better performance than the other models. Third, our model shows lower RMSEs for low-peak fluxes flares than those with high-peak fluxes flares. Fourth, our model shows high CC for flare duration prediction.

3.3. *Visual explanation of a deep learning solar flare forecast model*

We make a new DL flare forecasting model to visually show the DL model explanation and compare them with the physical parameters of solar active regions (Yi et al. 2021). For this, we use full-disk magnetograms from 1996 to 2017. The model forecasts "Yes or No" of daily flare occurrence for C-, M-, and X-class. For model explanation, we use CNN attribution methods, guided backpropagation and Gradient-weighted Class Activation Mapping (Grad-CAM). The major results of this study are as follows. First, our DL model successfully forecasts daily solar flare occurrence with TSS = 0.65. Second, by the attribution methods, we find that the flare model importantly considers active regions (Figure 14) and the polarity inversion line. Third, Grad-CAM values of ARs and their flare occurrence rates of them are highly correlated. Fourth, Grad-CAM values are well correlated with nine SHARP parameters such as total unsigned vertical current, total unsigned current helicity, total unsigned flux, and total photospheric magnetic free energy density.

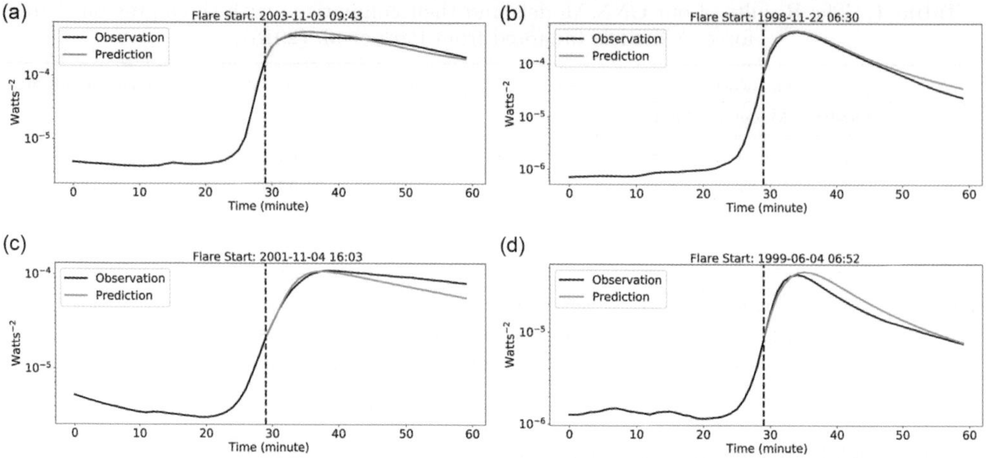

Figure 14. Solar flare X-ray forecasting results of X5 (a), X4 (b), X1 (c), and M4 (d) flare. (a), (b), (c), and (d) predictions are started 7, 7, 7, and 6 minutes after flare start time, respectively. The black vertical dashed lines denote forecast start times (adapted from Yi et al. (2020)).

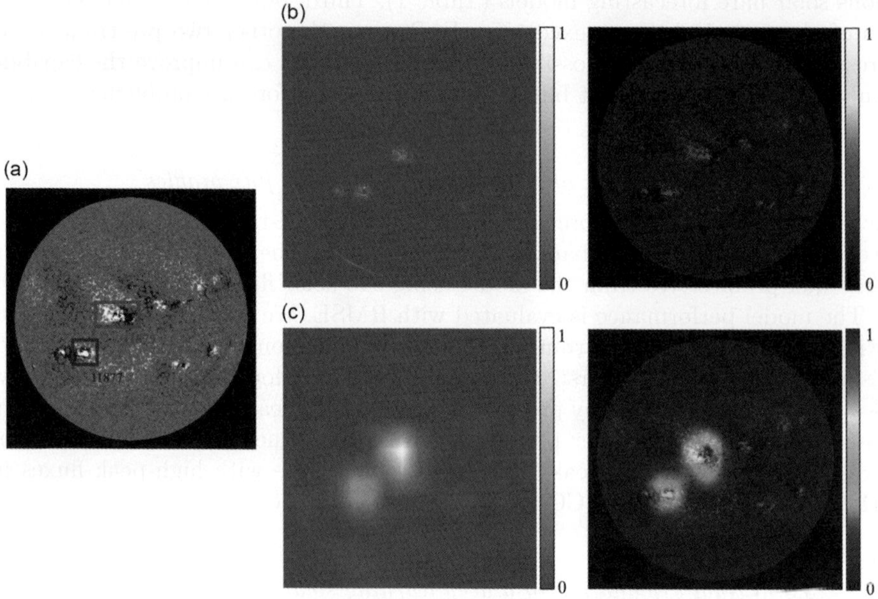

Figure 15. Results of attribution methods. (a) A line-of-sight magnetogram observed at 00:00 UT on 2013 October 23. (b) A guided backpropagation mask (left) and the mask overlaid on the magnetogram (right). (c) A Grad-CAM mask (left) and the mask overlaid on the magnetogram (right). The highlighted pixels/areas on masks indicate the gradient of pixels/areas for prediction (adapted from Yi et al. (2021)).

3.4. *Forecast of hourly relativistic (> 2 MeV) electron fluxes*

We forecast hourly relativistic (> 2 MeV) electron fluxes at geostationary orbit for the next 72 hours by using a multilayer perceptron based DL model (Son et al. 2022). The input data of our model are solar wind parameters: temperature, density, speed, interplanetary magnetic field (|B| and Bz), geomagnetic indices (Kp and Dst), and electron fluxes themselves. All input data are hourly averaged over the previous 72 hours. We obtain electron flux data from Geostationary Operational Environmental Satellite

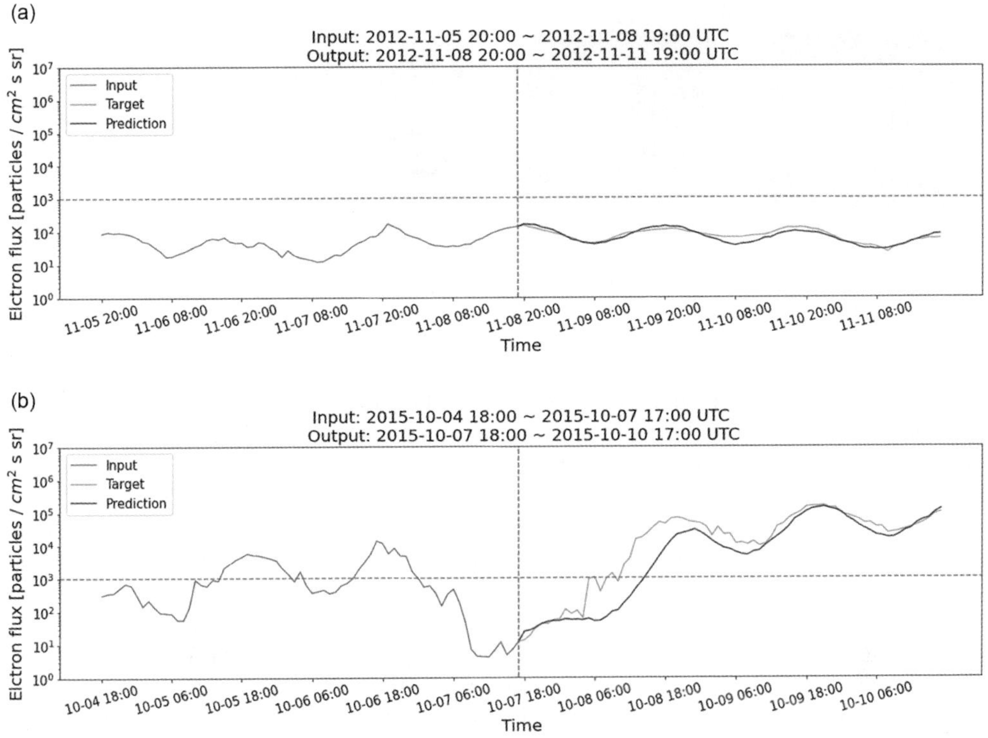

Figure 16. Two examples of our model results. Based on the red dashed line (prediction time), the left is the input time sequences and the right is the output time sequences. The gray lines are observed electron flux values during input time sequences, the yellow lines are target ones, and the blue lines are predicted ones by our model (adapted from Son et al. (2022)).

(GOES)-15 and -16, and perform the calibration for matching these two data. The total period of the data is from 2011 January to 2021 March (GOES-15 data for 2011–2017 and GOES-16 data for 2018–2021). We divide the dataset into a training set (January-August), a validation set (September), and a test set (October-December) to consider the solar cycle effect. Our major results are as follows. First, our model successfully predicts hourly electron fluxes for the next 72 hours (Figure 16). Second, prediction efficiency is from 0.97 (for 1 hr prediction) to 0.53 (for 72 hr prediction), and the root-mean-square error of our model is from 0.18 to 0.68, which are much better than those of the previous studies. Third, our model well predicts both diurnal variation and sudden changes in electron fluxes associated with fast solar winds and interplanetary magnetic fields. Our results demonstrate that the DL model can be applied to forecasting long-term sequential space weather events.

3.5. *Forecast of 1-day TEC maps*

We present a global TEC forecasting model using a novel DL method based on conditional generative adversarial networks (Yi et al. 2021). For training, we use the International GNSS Service (IGS) TEC maps from 2003 to 2012 with 2-hour time cadence. Our model has two input images (IGS TEC map and 1-day difference map between the present day and the previous day) and one output image (1-day future map). The model is tested with two different data sets: solar maximum period (2013–2014) and solar minimum period (2017–2018). Then, we compare the results of our model

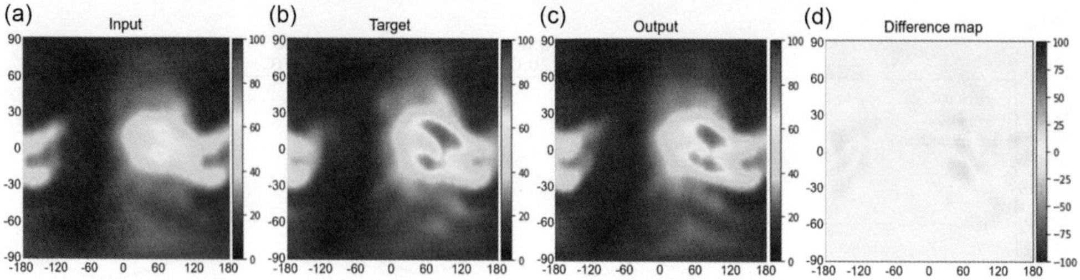

Figure 17. Comparison between IGS TEC maps and the results of our model at 02 UT on October 1, 2013. (a) input image, (b) target image (+24h IGS TEC), (c) output image generated by our model, (d) difference map (between [b and c]) (adapted from Lee et al. (2021b)).

with those of the 1-day CODE (Center for Orbit Determination in Europe, Schaer 1999) prediction model. Our major results can be summarized as follows. First, we successfully apply our model to the forecast of global TEC maps. Second, our model well predicts daily TEC maps with 1 day in advance using only previous TEC maps (Figure 17). The averaged root mean square error, bias, and standard deviation between AI-generated and IGS TEC maps are 2.74 TECU, -0.32 TECU, and 2.59 TECU, respectively. Third, our model generates some peak structures around equatorial regions. Fourth, our model shows better results than the 1-day CODE prediction model during both solar maximum and minimum periods. Fifth, another model with additional input data Kp index gives a slight improvement in the results. Our study demonstrates that our DL model based on an image translation method will be effective for forecasting future images using previous data.

4. Prospects

We have successfully applied novel DL methods to solar and space weather data. Especially, the task of image-to-image translation is quite promising and our new method (Pix2PixCC) is very useful for many scientific data. We are extending our applications to several cases: EUV images from historic Ca II ones, EUV images from coronal white light ones, H-alpha images from EUV ones, pixel-to-pixel translation among EUV images, fast construction of coronal physical parameters based on MAS, fast construction of coronal force-free fields, and three days forecast of global TEC maps.

In addition, the space weather applications by DL are underway in various aspects: application of reinforcement learning to solar flare forecast, construction of the ionospheric 3-D density model, automatic determination of CME parameters (like CDAW, Gopalswamy et al. 2009), automatic determination of CME 3-D parameters, and forecast of solar wind data, kp and Dst index using solar and interplanetary data. We have a plan to set up an integrated space weather model based on DL by combining the data from several different components: solar data, solar explosive events, solar wind, magnetosphere, and ionosphere/thermosphere.

Acknowledgements

We are very thankful to several teams for providing space-based and ground-based observation data: SOHO, SDO, STEREO, GOES, ACE, etc.

References

Bilitza, D., Altadill, D., Truhlik, V., et al. 2017, Space Weather, 15, 418
Bloomfield, D. S., Higgins, P. A., McAteer, R. T. J., & Gallagher, P. T. 2012, ApJ Letters, 747, L41

Colak, T. & Qahwaji, R. 2009, Space Weather, 7, S06001
de Wijn, A. G., Burkepile, J. T., Tomczyk, S., et al. 2012, in Society of Photo-Optical Instrumentation Engineers (SPIE) Conference Series, Vol. 8444, Ground-based and Airborne Telescopes IV, ed. L. M. Stepp, R. Gilmozzi, & H. J. Hall, 84443N
Domingo, V., Fleck, B., & Poland, A. I. 1995, Solar Physics, 162, 1
Ermolli, I., Criscuoli, S., Centrone, M., Giorgi, F., & Penza, V. 2007, A&A, 465, 305
Ermolli, I., Fofi, M., Bernacchia, C., et al. 1998, Solar Physics, 177, 1
Galilei, G., Welser, M., & de Filiis, A. 1613, Istoria E dimostrazioni intorno alle macchie solari E loro accidenti comprese in tre lettere scritte all'illvstrissimo signor Marco Velseri ...
Gopalswamy, N., Yashiro, S., Michalek, G., et al. 2009, Earth Moon and Planets, 104, 295
Hochreiter, S. & Schmidhuber, J. 1997, Neural Comput., 9, 1735–1780
Huang, X., Wang, H., Xu, L., et al. 2018, ApJ, 856, 7
Isola, P., Zhu, J.-Y., Zhou, T., & Efros, A. A. 2017, in Proceedings of the IEEE Conference on Computer Vision and Pattern Recognition (CVPR)
Jeong, H.-J., Moon, Y.-J., Park, E., & Lee, H. 2020, ApJ Letters, 903, L25
Jeong, H.-J., Moon, Y.-J., Park, E., Lee, H., & Baek, J.-H. 2022, The Astrophysical Journal Supplement Series, 262, 50
Ji, E.-Y., Moon, Y.-J., & Park, E. 2020, Space Weather, 18, e02411
Kaiser, M. L., Kucera, T. A., Davila, J. M., et al. 2008, Space Science Reviews, 136, 5
Keller, C. U., Harvey, J. W., & Giampapa, M. S. 2003, in Innovative telescopes and instrumentation for solar astrophysics, Vol. 4853, International Society for Optics and Photonics, 194–204
Kim, T., Park, E., Lee, H., et al. 2019, Nature Astronomy, 3, 397
Kumar, S., Tan, E. L., & Murti, D. S. 2015, Earth, Planets and Space, 67, 42
Lawrance, B., Lee, H., Park, E., et al. 2022, ApJ, 937, 111
Lee, H., Park, E., & Moon, Y.-J. 2021a, ApJ, 907, 118
Lee, S., Ji, E.-Y., Moon, Y.-J., & Park, E. 2021b, Space Weather, 19, e2020SW002600
Lemen, J. R., Title, A. M., Akin, D. J., et al. 2012, Solar Physics, 275, 17
Lim, D., Moon, Y.-J., Park, E., & Lee, J.-Y. 2021, ApJ Letters, 915, L31
Mirza, M. & Osindero, S. 2014, arXiv preprint arXiv:1411.1784
Park, E., Moon, Y.-J., Lim, D., & Lee, H. 2020, ApJ Letters, 891, L4
Park, E., Moon, Y.-J., Shin, S., et al. 2018, ApJ, 869, 91
Pesnell, W. D., Thompson, B. J., & Chamberlin, P. C. 2012, Solar Physics, 275, 3
Rahman, S., Moon, Y.-J., Park, E., et al. 2020, ApJ Letters, 897, L32
Schaer, S. 1999, Geod.-Geophys. Arb. Schweiz, 59
Scherrer, P. H., Bogart, R. S., Bush, R. I., et al. 1995, Solar Physics, 162, 129
Schou, J., Scherrer, P. H., Bush, R. I., et al. 2012, Solar Physics, 275, 229
Selvaraju, R. R., Cogswell, M., Das, A., et al. 2017, in 2017 IEEE International Conference on Computer Vision (ICCV), 618–626
Shin, G., Moon, Y.-J., Park, E., et al. 2020, ApJ Letters, 895, L16
Son, J., Cha, J., Moon, Y.-J., et al. 2021, ApJ, 920, 101
Son, J., Moon, Y.-J., & Shin, S. 2022, Space Weather, 20, e2022SW003153
Springenberg, J. T., Dosovitskiy, A., Brox, T., & Riedmiller, M. 2014, Striving for Simplicity: The All Convolutional Net
Sutskever, I., Vinyals, O., & Le, Q. V. 2014, Sequence to Sequence Learning with Neural Networks
Wang, T.-C., Liu, M.-Y., Zhu, J.-Y., et al. 2018, in 2018 IEEE/CVF Conference on Computer Vision and Pattern Recognition, 8798–8807
Yi, K., Moon, Y.-J., Lim, D., Park, E., & Lee, H. 2021, ApJ, 910, 8
Yi, K., Moon, Y.-J., Shin, G., & Lim, D. 2020, ApJ Letters, 890, L5

The Era of Multi-Messenger Solar Physics
Proceedings IAU Symposium No. 372, 2023
G. Cauzzi & A. Tritschler, eds.
doi:10.1017/S1743921322004859

Image refinement and estimations of radiation formation heights with the Deep Solar ALMA Neural Network Estimator

Henrik Eklund

Institute for Solar Physics, Department of Astronomy, Stockholm University AlbaNova University Centre, SE-106 91 Stockholm, Sweden

Abstract. The signatures of small-scale features in the solar atmosphere are severely degraded by limited angular resolution of the observations. The Deep Solar ALMA Neural Network Estimator (Deep-SANNE) is trained towards synthetic observables from 3D magnetohydrodynamic simulations to recognize the small-scale dynamic features in data at limited observational resolution, and provide maps of correction factors across the field of view. The correction factors can be used to acquire deconvolved refined images with significantly improved brightness temperature contrasts, where the strength of brightening events are reproduced to an accuracy of 94.0% instead of the 43.7% at observational resolution. Deep-SANNE can also provide masks of the most probable locations with large accuracies, and estimations on the radiation formation heights in connection to the small-scale features. The Deep-SANNE refined images and estimations of radiation formation heights allow for larger accuracy and meaningful analysis of solar ALMA data.

Keywords. techniques: high angular resolution, techniques: image processing, techniques: interferometric, instrumentation: high angular resolution, instrumentation: interferometers, Sun: atmospheric motions, Sun: radio radiation, Sun: chromosphere, shock waves, stellar dynamics

1. Introduction

There are a lot of dynamic features at small scales in the solar chromosphere, which are important to study for the understanding of the underlying processes. The angular resolution of observations however limits the observability of the small-scale features. The resolution scales with wavelength and inversely with the size of the aperture. Therefore interferometry is useful to achieve a high angular resolution in the radio regime. With aperture sizes on the order of km, the Atacama Large Millimeter/sub-millimeter Array (ALMA) offers solar observations at novel sub-arcsec resolution. For now (in cycle 9; 2022–2023), ALMA offers solar observations at four receiver bands (Table 1) between 100.0 GHz (3.0 mm) and 346.6 GHz (0.87 mm), with the highest angular resolution of 0.62 arcsec at 239.0 GHz (1.25 mm).

There are however features in the solar chromosphere at scales of the same order or smaller than this novel high resolution. As a result, the strength of small-scale dynamic brightening events in solar ALMA data originating from e.g., propagating shock waves appears lesser than indicated by numerical simulations (Eklund et al. 2020, 2021a). A systematic study was deployed by Eklund et al. (2021b), to detect brightening events in 3D magneto-hydrodynamic Bifrost simulations (Gudiksen et al. 2011; Carlsson et al. 2016) and derive correction factors for the degradation originating from the spatial resolution of the different ALMA receiver bands and resolutions. However as these correction factors were based on a statistical sample with many events being differently well resolved,

Table 1. ALMA receiver bands with their centre frequencies, centre wavelengths, their largest inteferometric array configuration and corresponding highest angular resolution offered for solar observations in cycle 9 (2022-2023). (The values of the maximum angular resolutions given here assume the use of Briggs weighting with robust parameter of 0.5 (Privon et al. 2022)).

Receiver band	Frequency [GHz]	Wavelength [mm]	Array configuration	Max. angular resolution [arcsec]
Band 3	100.0	3.0	C4	0.92
Band 5	198.0	1.58	C3	0.77
Band 6	239.0	1.25	C3	0.62
Band 7	346.6	0.87	C2	0.67

they show significant uncertainties. This motivated the study and development of the Deep Solar ALMA Neural Network Estimator (Deep-SANNE) to distinguish how well resolved each individual event is and to what degree it is degraded. Using an artificial neural network that analyses the combined temporal and spatial dynamics of the observed small-scale features, not only the degradation of brightening events can be retrieved, but also maps of correction factors across the entire field of view. This allows for studying all type of small-scale events at improved contrasts. Based on simulations, such an artificial neural network is also capable to correlate the spatio-temporal small-scale features to e.g., the radiation height of formation.

The applications here are focused on the case of band 6 with the highest resolution offered for solar observations (Table 1), with a realistic slightly elliptic point spread function. An overview of the architecture and training procedure of the neural network is given below followed by a couple examples of applications showing the refined images and estimations of the radiation formation heights. See also '*Deep Solar ALMA Neural Network Estimator for image refinement and estimations of small scale dynamics*' by Eklund (2022) and the Deep-SANNE git repository†, for more examples and details.

2. Architectural design and training of the neural network

The Deep Solar ALMA Neural Network Estimator (Deep-SANNE) is a spatio-temporal artificial neural network based on a combination of convolutional Long-Short Term Memory (LSTM; Hochreiter & Schmidhuber 1997; Shi et al. 2015) units and densely connected layers. A LSTM neural network is a type of recurrent neural network (RNN; Rumelhart et al. 1986) that through a series of gates is specialized to handle long sequences of data and distinguish (and remember) the essential information from both the beginning and the end of the sequence.

The architecture of Deep-SANNE consists of a series of layers and is illustrated in Fig. 1. Deep-SANNE takes the input of a time-series of length i brightness temperature images at observational angular resolution, with spatial dimensions $n \times m$. The second layer is a densely connected layer, with dimensions $n \times m \times l_1$. The third layer consists of a series of LSTM cells, each with dimensions $n \times m \times l_2$, where two-dimensional convolutional operations are performed at the gates regulating the flow of information through the network (Shi et al. 2015). The forth layer is another densely connected layer with dimensions $n \times m \times l_3$ and finally the output layer providing an image of dimensions $n \times m$ for the last time-step t^i. Further details on the architecture of Deep-SANNE are described by Eklund (2022). The densely connected layers and the LSTM cells each had depths of $l_1 = 64$ $l_2 = 128$ and $l_3 = 128$, respectively. Time-series of length 100 s were used with $i = 24$ frames, with a cadence of 10 s between $t = 0 - 60$ s and 2 s cadence between $t = 68 - 100$ s. For the current applications, Deep-SANNE is trained to 2 s cadence to

† `https://github.com/henrikeklund/Deep-SANNE/`

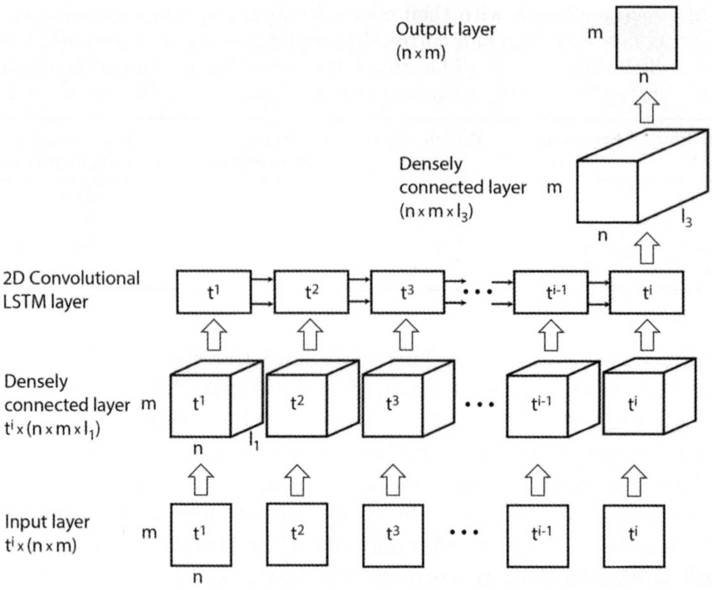

Figure 1. Schematics of the architecture of Deep-SANNE. Figure adapted from Eklund (2022).

be compatible with the early solar ALMA observational datasets. However, it is fully compatible to be applied to datasets with higher cadence, e.g., 1 s.

Two different numerical simulations from the code Bifrost (Gudiksen et al. 2011) were used for training and validating Deep-SANNE. The first is a high candence (1 s) re-run of the publicly available enhanced network (EN) simulation (Carlsson et al. 2016) that exhibits quiet Sun with network patches, with approximately 1 hour length. The second is an additional simulation also exhibiting quiet Sun but without any large scale magnetic field network features, resebmling the conditions of a coronal hole (CH; a snapshot of synthetic observables is shown in Fig. 2a), with a cadence of 2 s and a length of only 240 s. Both simulations include treatment of non-local thermodynamic equilibrium and non-equilibrium hydrogen ionization. The observables at the ALMA wavelengths were acquired by calculations with the Advanced Radiative Transfer code (de la Cruz Rodríguez et al. 2021). The synthetic observable maps were then also degraded to the resolution of the observations. About 31000 input sequences were used for training the neural network and the data used for validation was never seen by the neural network.

3. Deconvolution and Deep-SANNE refined images

Given a time-series of images with brightness temperatures at observational resolution (Fig. 2b being the last frame, t^i), Deep-SANNE gives output of maps of correction factors with the estimated degradation that comes with the limited angular resolution, relative to the highly resolved original model (see example in Fig.3e of Eklund 2022). These maps with correction factors can be applied to the data at observational resolution to acquire the Deep-SANNE refined image (Fig. 2c). The contrasts and span of the brightness temperatures are significantly improved in the refined image and more small-scale features are revealed that are not apparent in the image at observational resolution. Deep-SANNE thus functions as a temporal-spatial deconvolver, that by using the added information of the temporal domain achieves better than a standard 2D deconvolver, such

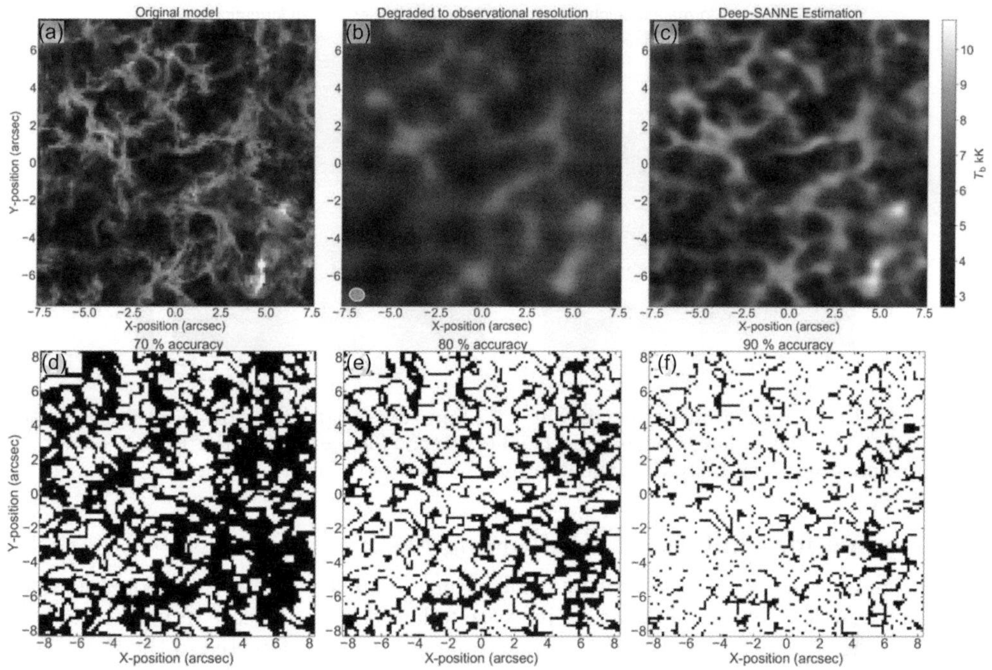

Figure 2. Deep-SANNE deconvolution and accuracy indications. (a) Brightness temperatures at 1.25 mm of a CH simulation snapshot. (b) The same snapshot downgraded to angular resolution corresponding to observations. (c) The corresponding Deep-SANNE refined image of the same snapshot. (d-f) Masks from Deep-SANNE indicating locations where the brightness temperatures at observational resolution are at least 70%, 80% and 90% accurate. Figure adapted from Eklund (2022).

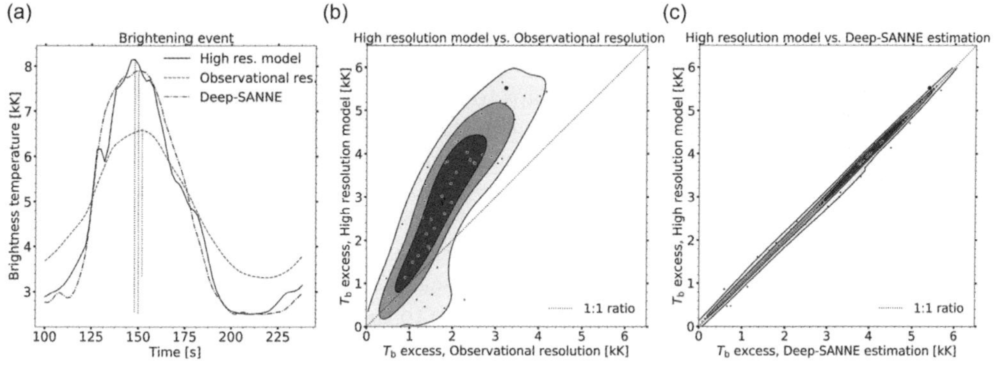

Figure 3. Brightening events in the Bifrost model. (a) Example of a brightening event at observational and high resolution and at the Deep-SANNE refined images. (b) Distribution of strengths at original resolution against observational resolution. (c) Distribution of strengths at original resolution against Deep-SANNE refined images. Panels adapted from Eklund (2022).

as the Richardson-Lucy algorithm (Eklund 2022). The output maps of correction factors from Deep-SANNE can also be used to construct masks indicating at what locations in the field of view the brightness temperatures at observational resolution are probable to have a large accuracy. Figs. 2d-f show the masks of locations that have at least 70%, 80% and 90% accuracy. These type of masks could be very useful for selecting reliable locations for eventual in-depth analysis of observational data.

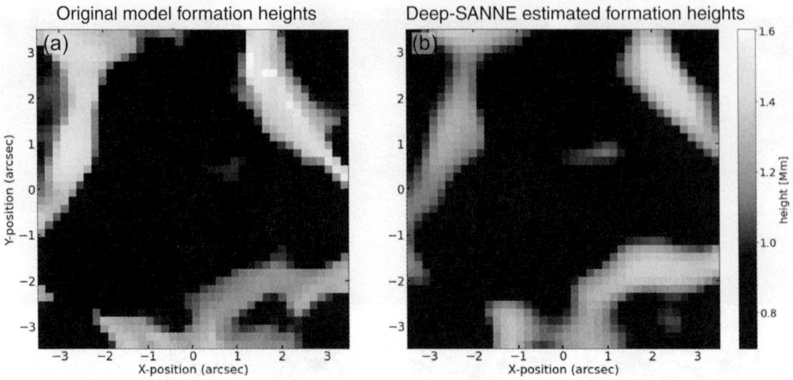

Figure 4. Radiation formation heights at 1.25 mm wavelength (ALMA band 6), at optical depth unity. (a) The radiative formation heights of the calculations on the Bifrost model. (b) The corresponding Deep-SANNE estimations of the radiation formation height.

4. Example for analysis of brightening events

The refined images are useful for analysis of any small-scale feature or dynamic event. Here an example of events showing an increase followed by a decrease in brightness temperature (Fig. 3a) is shown. About 53 brightening events were detected in the CH dataset (Eklund 2022), and their strength at the original high resolution were compared to that seen in the degraded observational resolution (Fig. 3b) and in the Deep-SANNE refined images (Fig. 3c). While the events at observational resolution show a broad distribution, originating from more or lesser resolved events, the Deep-SANNE refined images show a strongly linear correlation close to the 1:1 ratio of the original high resolution, successfully distinguishing how well resolved the individual events are.

5. Estimation of millimeter wavelength radiation formation heights

Deep-SANNE was also trained to correlate the spatio-temporal small-scale dynamic features in the brightness temperature maps to the radiation formation heights (at optical depth unity) provided from the radiative transfer calculations. An example of the radiation formation heights from the radiative transfer calculations on the simulation are given in Fig. 4a and the corresponding Deep-SANNE estimations are given in Fig. 4b. A lot of the small-scale features in the formation height map (Fig. 4a) are recovered by the estimations of Deep-SANNE (Fig. 4b). There are however some features that are not well recovered, originating from features at scales well below the resolution element. Indication of the height of formation in connection to the small scale dynamic at different locations in the field of view is useful for the interpretation of the observational data. The estimations of the formation heights are model dependent and awareness on the specifics of the simulation and radiative transfer calculations needs to be taken into account when performing analysis of the estimated formation heigts.

6. Slope of brightness temperature continuum

Instead of reconstructing images with brightness temperatures using the whole receiver band (Table 1), individual images could be reconstructed for different sub-bands. There are significant differences in brightness temperature at wavelengths at opposite sides of the ALMA receiver bands (e.g., at 1.2 mm and 1.3 mm), which provides additional information and a measurement of the gradient of the plasma temperature at the probed layers in the solar atmosphere (Eklund et al. 2022). Deep-SANNE could be trained and applied to time-series of sub-band images to improve the accuracy of studies making use of differences withing the spectral domain of ALMA observations.

7. Conclusions

An artificial spatio-temporal deep neural network, Deep-SANNE, is trained to recognize the dynamic of the of small-scale features in data at observational resolution. The output from Deep-SANNE can be used as (*i*) a deconvolver technique to acquire refined images with improved contrasts revealing more of the small-scale features. The spatio-temporal deconvolver achieves better than a standard 2D spatial deconvolver, such as the Richardson-Lucy algorithm (Eklund 2022), (*ii*) acquire estimations of the accuracy of brightness temperatures across the field of view and (*iii*) perform estimations of the radiation formation heights in connection to the small-scale features. The applications of Deep-SANNE can improve the analysis of small-scale features and allow for more meaningful interpretation of solar ALMA data.

Acknowledgements

This work was supported through the CHROMATIC project (2016.0019) funded by the Knut and Alice Wallenberg foundation and the Research Council of Norway through its Centres of Excellence scheme, project number 262622, and through grants of computing time from the Programme for Supercomputing.

References

Carlsson, M., Hansteen, V. H., Gudiksen, B. V., Leenaarts, J., & De Pontieu, B. 2016, A&A, 585, A4

de la Cruz Rodríguez, J., Szydlarski, M., & Wedemeyer, S. 2021, ART: Advanced (and fast!) Radiative Transfer code for Solar Physics (https://github.com/SolarAlma/ART).

Eklund, H. 2022, A&A, Deep Solar ALMA Neural Network Estimator for image refinement and estimations of small scale dynamics, in review.

Eklund, H., Szydlarski, M., & Wedemeyer, S. 2022, A&A

Eklund, H., Wedemeyer, S., Snow, B., et al. 2021a, Philosophical Transactions of the Royal Society of London Series A, 379, 20200185

Eklund, H., Wedemeyer, S., Szydlarski, M., & Jafarzadeh, S. 2021b, A&A, 656, A68

Eklund, H., Wedemeyer, S., Szydlarski, M., Jafarzadeh, S., & Guevara Gómez, J. C. 2020, A&A, 644, A152

Gudiksen, B. V., Carlsson, M., Hansteen, V. H., et al. 2011, A&A, 531, A154

Hochreiter, S. & Schmidhuber, J. 1997, Neural computation; MIT Press, 9, 1735

Privon, G., Nagai, H., Rebolledo, D., & Díaz Trigo, M. 2022, ALMA Cycle 9 Proposer's Guide, ALMA Doc. 9.2 v1.4, Vol. 9.2

Rumelhart, D. E., Hinton, G. E., & Williams, R. J. 1986, nature, 323, 533

Shi, X., Chen, Z., Wang, H., et al. 2015, arXiv e-prints, arXiv:1506.04214

Henrik Eklund of Stockholm University (Sweden) asking a question during the session on Machine Learning.

IAU Division E Advisor Sarah Gibson (HAO, USA) and President Cristina Mandrini (CONICET, Argentina), during the Symposium S372.

A group photo of IAU participants taken on August 8, 2022. Alexandra Tritschler and Gianna Cauzzi, IAU S372 co-Chairs, are 4th and 5th from left, respectively, in the top row of Zoom participants.

Author index